Disappearing Destinations: Climate Change and Future Challenges for Coastal Tourism

Disappearing Destinations: Climate Change and Future Challenges for Coastal Tourism

Editor

Kornelia Marshall

scitus
academics

Disappearing Destinations: Climate Change and Future Challenges for Coastal Tourism

Edited by **Kornelia Marshall**

ISBN: 978-1-68117-251-4
Library of Congress Control Number: 2016934794

SCITUS Academics LLC,
www.scitusacademics.com
Box No. 4766, 616 Corporate Way,
Suite 2, Valley Cottage,
NY 10989

Notice

Reasonable efforts have been made to publish reliable data and views articulated in the chapters are those of the individual contributors, and not necessarily those of the editors or publishers. Editors or publishers are not responsible for the accuracy of the information in the published chapters or consequences of their use. The publisher believes no responsibility for any damage or grievance to the persons or property arising out of the use of any materials, instructions, methods or thoughts in the book. The editors and the publisher have attempted to trace the copyright holders of all material reproduced in this publication and apologize to copyright holders if permission has not been obtained. If any copyright holder has not been acknowledged, please write to us so we may rectify.

Preface

A tourist destination is a city, town, or other area that is dependent to a significant extent on the revenues accruing from tourism, or "a country, state, region, city, or town which is marketed or markets itself as a place for tourists to visit". It may contain one or more tourist attractions and possibly some "tourist traps.Many of the world's most spectacular landscapes are vanishing because of climate change, spurring concerned visitors to experience and protect them before it's too late. What would an autumn cycling trip in New England be without colorful maples? What's a ski vacation without fresh snow? Or an outing to the shore where the beach has eroded? These scenarios are unimaginable for many, yet global warming threatens to make them a reality as species extinctions, severe storms, flooding, drought, melting icecaps, and warmer, more acidic ocean water transform the outdoor environments we love. Vanishing destinations such as the Great Barrier Reef, Maldives, Galapagos Island and polar regions provide tour operators with an opportunity to market their products with the unique selling point 'to see it before it is gone' tourism. Bleaching corals, melting ice caps disappearing islands and wildlife are indicators that put these destinations on the 'must-see' lists. This kind of tourism defined as' Doom tourism', 'Climate tourism' or 'Last-chance tourism, which is the preferred term by academics, is a trend whereby tourists more and more seek to experience most endangered sites in the world, before they will be gone. Visiting disappearing destinations can provide unique opportunities to develop environmental awareness. Furthermore the increasing demand to these specific destinations is economically beneficial for tour operators and the local community, who may even depend on tourism for their livelihood.This book,Disappearing Destinations: Climate Change and Future Challenges for Coastal Tourism, discusses threats and consequences of climate predictions on coastal tourism destinations. It will be an essential tool for researchers, policymakers and students in tourism, ecotourism, environmental conservation, planning, coastal management and engineering, climate change and marine conservation.

Table of Contents

Chapter 1

Convinced, Ambivalent or Annoyed: Tyrolean Ski Tourism Stakeholders and Their Perceptions of Climate Change

Lisa Trawöger'

Institute of Geography, University of Innsbruck, Innrain 52, 6020 Innsbruck, Austria

ABSTRACT

Its focus on snow-dependent activities makes Alpine winter tourism especially sensitive to climate change. Stakeholder risk perceptions are a key factor in adaptation to climate change because they fundamentally drive or constrain stakeholder action. This paper examines climate change perceptions of winter tourism stakeholders in Tyrol (Austria). Using a qualitative approach, expert interviews were conducted. Four opinion categories reflecting different attitudes toward climate change issues were identified: *convinced planners, annoyed deniers, ambivalent optimists, convinced wait-and-seers*. Although the findings generally indicate a growing awareness of climate change, this awareness is mainly limited to perceiving the issue as a global phenomenon. Awareness of regional and branch-specific consequences of climate change that lead to a demand for action could not be identified. Current technical strategies, like snowmaking, are not primarily climate-induced. At present, coping with climate change is not a priority for risk management. The findings point out the importance of gaining and transferring knowledge of regional and branch-specific consequences of climate change in order to induce action at the destination level.

Keywords: Climate change impacts; Snow-based winter tourism; Ski tourism; Stakeholder; Risk perception; Tyrol; Austria

1. INTRODUCTION

Climate change is increasingly regarded as the single most serious global environmental problem (Scott, Gössling, & Hall, 2012; see also Houghton, 2009). Significant variations in the mean state or variability of climate are major drivers of global environmental change encompassing changes in ocean currents, land cover or biodiversity. Environmental change on a global scale will result in possibly far-reaching negative effects at the regional or even local level, ranging from problems with water availability and water quality to soil erosion and flooding (Houghton, 2009 and IPCC, 2007). Because of strong interlinkages, global environmental change is in turn impacting climate change.

Modified temperature and precipitation patterns will also have considerable consequences for various socio-economic systems, including tourism. Since climate is a key resource for several forms of tourism, climate change is likely to pose a major challenge for many destinations (Scott, Amelung, et al., 2008).

International tourism is a global key driver of socio-economic progress; it has a central role in creating jobs and in providing infrastructure (United Nations World Tourism Organization (UNWTO), 2013). Tourism is also of vital importance to the European economy. A total of 534 million international tourist arrivals represent a global market share of 52% for European tourism (UNWTO, 2013). Tourism accounts for more than 5% directly and about 10% indirectly of the EU GDP (European Commission, 2012). European tourism is especially intense in coastal regions, some cities and Alpine regions (Eurostat, 2011).

According to Becken and Hay (2007), the European Alps generate about 7–10% of the annual global income from all-year tourism, with some 100 million tourists visiting the Alps each year. Many mountain and winter tourism destinations in the European Alps are among the most frequently visited regions in the world. Many forms of Alpine winter tourism are dependent on snow-based activities. In many rural areas, winter tourism has become the primary source of income (Abegg, Agrawala, Crick, & de Montfalcon, 2007).

Since the 1970s, downhill skiing has expanded heavily in the over 600 ski resorts located in France, Switzerland, Austria and Italy (Abegg et al., 2007). The number of domestic and international skier visits to the European Alps shows the dominance of this region in worldwide winter tourism. While 400 million skier visits are generated globally, the Alps account for 176 million skier visits per year, representing 44% of the total visits (Vanat, 2013). Austria is among the largest national markets, generating about 51 million skier visits per year (WKO, 2012).

Its focus on snow-dependent activities makes Alpine winter tourism highly vulnerable to the effects of an ongoing warming trend (Scott, 2011 and Scott et al., 2012). Consequently, regional climate change projections for Europe and the Alps are essential for assessing the future of snow-based tourism. Kotlarski, Bosshard, Fischer, Lüthi, and Schär (2010) compared current regional climate change scenarios for Europe and the Alps. The greenhouse gas forcing used by these models was derived from the Intergovernmental Panel on Climate Change (IPCC) SRES A1B emission scenario (IPCC, 2000). Depending on the applied model chain and season, winter temperatures are projected to increase by 0.8–2.6 °C until 2050 and by 2.9–4.2 °C until 2100, compared to the 1961–1990 baseline. Precipitation changes were found to be highly uncertain until 2050. However, by the end of the twenty-first century, there is model consensus regarding drier summer seasons and wetter winter seasons (an increase of 6.7–19.7% compared to baseline climate). Moreover, Kotlarski et al. (2010) report a decrease in snow cover resulting in an almost complete loss of snow in lowlands by 2100.

In a study published by the Organisation for Economic Co-operation and Development (OECD), Abegg et al. (2007) found a decline in the number of naturally snow-reliable Alpine ski areas from 609 to 404 (under a +2 °C warming scenario) and to 202 (under a +4 °C warming scenario). The authors concluded that climate change is a serious challenge to the Alpine ski industry, which was confirmed by Steiger and Abegg (2011) for Austrian ski areas. In the latter study, a site-specific physical snow module was applied, also taking snowmaking into account. Steiger and Abegg (2011) showed that the impact of climate change on Austrian ski areas can be balanced with snowmaking under a warming scenario of 1–2 °C. Assuming a warming of more than 2 °C, snowmaking is not considered to be an appropriate strategy because operational costs would increase dramatically.

The severity of effects in the respective tourism regions and destinations will heavily depend on their individual ability to adapt to climate change. Adaptation in the context of climate change refers to "adjustments in ecological-socio-economic systems in response to actual or expected climatic stimuli, their effects or impacts" (Smit, Burton, Klein, & Wandel, 2000, p. 225). Adaptations can be anticipatory (taken proactively before the climate stimulus), concurrent (during) or reactive (in response). Based on their degree of purposefulness, Smit et al. (2000) distinguish between autonomous (spontaneous) and planned (deliberate) adaptive responses to an actual or anticipated stimulus. Adaptations in socio-economic systems are often concurrent or reactive in nature (Smit et al., 2000).

There are many different forms of adaptive responses to climate change. Smit et al. (2000) distinguish between primarily technological, behavioural, financial, institutional and informational adaptations, while Scott et al. (2012) provide a more detailed classification scheme of technical, economic, policy, institutional, managerial, planning, legal and behavioural climate adaptations. Adaptive strategies in the snow-based tourism industry range from technical actions (e.g., increased use of snowmaking) to managerial responses, including the introduction of alternative tourism products to skiing in the winter season or low-season closures (Dawson and Scott, 2013, Hall and Higham, 2005, Scott, Dawson, et al., 2008, Scott et al., 2009 and Scott et al., 2012). More recent strategies also involve behavioural action, such as the use of real-time webcams of snow conditions (Scott et al., 2012).

Effective adaptation reduces a region's vulnerability to climate change. On the local or regional level, the potential of a system to adapt is not only determined by economic and technical resources but also by such factors as managerial ability or access to information resources (Smit & Wandel, 2006). Stakeholder perceptions are another crucial aspect in reducing vulnerability to climate change. The key players' awareness of climate change determines whether effective action is taken (Belle and Bramwell, 2005 and Klint et al., 2012). Stakeholder estimations of the possible scale and effects of the threat are decisive for the perceived need to adapt in a region. According to Belle and Bramwell (2005), perceptions of individuals and interest groups are also crucial to both the development of public policies and the acceptance of proposed policy measures. Correspondingly, learning to live with change and uncertainty is due not least of all to how well stakeholders in a region perceive potential challenges. In the context of climate change research, risk perception studies among stakeholders are vitally important in assessing a region's vulnerability to climate change.

Extensive research has been conducted on the relationship between snow-based tourism and climate change in recent years (e.g. Hall, 2006, Hall, 2011, Scott, Dawson, et al., 2008, Scott and McBoyle, 2007, Scott et al., 2007, Steiger, 2010 and Steiger, 2011). However, empirical social research on the supply side, focusing on the perception of climate change among key players in the ski industry, is still limited.

The present study aims to fill that gap in one of the most important European snow-based winter tourism regions. More specifically, the study (1) reveals the level of climate change awareness of and involvement with the issue in one of the top Alpine tourism regions; (2) identifies the present role of climate change in regional and destination risk management; (3) elaborates on the significance of snowmaking for the ski industry; and thus (4) contributes to the literature on supply-side perception in the context of climate change and snow-based winter tourism.

2. LITERATURE REVIEW

2.1. Risk perception research

Risk perception research is rooted in the nuclear debate of the 1960s (Sjöberg, 2000) and has been of particular interest to scientists and policymakers since that period. The fields of psychology, sociology, political science, geography and anthropology have contributed to the early literature pertaining to this issue (Slovic, 1987). Several theoretical approaches have been developed to explain differences in risk perception among individuals and groups (e.g. Douglas and Wildavsky, 1982, Fischhoff et al., 1978 and Kasperson et al., 1988).

Research (e.g. Slovic, 1987 and Slovic, 2000) has shown that perceived risk is not only influenced by technical risk estimates but also determined by several qualitative characteristics related to whether a hazard is catastrophic or chronic, dreaded or common, old or new and delayed or immediate. The extent to whether a hazard is controllable, observable, fatal and known to science also shapes risk perceptions. Moreover, various social and cultural factors influencing risk perception (e.g. Baron, 2006, Berger et al., 2011, Costa Font et al., 2009, Sjöberg, 2000, Sunstein, 2006 and Wachinger et al., 2012) and a range of psychological barriers to addressing climate change (Gifford, 2011) have been identified.

Studies of climate change risk perception examine participants' awareness of possible effects of climate change. Recent research has found a discrepancy between the scientific community and the general public with regard to perceptions of challenges posed by climate change (e.g. Etkin and Ho, 2007 and Lorenzoni and Pidgeon, 2006). The complex, pervasive and time-delayed nature of climate change risks was found to lead to a limited understanding of possible consequences among lay people (e.g. Lorenzoni and Pidgeon, 2006 and Weber, 2006).

Climate change perception research in the field of tourism not only highlights barriers to adaptation and reveals knowledge gaps but also indicates opportunities for stakeholder action (Klint et al., 2012). A growing body of literature has investigated tourist and stakeholder climate change perceptions in various types of tourism destinations, including coastal tourism (e.g. Braun et al., 1999, Buckley, 2008, Buzinde et al., 2010, McEvoy et al., 2008, Rauken et al., 2010 and Valls and Sardá, 2009), island tourism (e.g. Becken, 2005, Belle and Bramwell, 2005, Gössling et al., 2006, Hall and Clayton, 2009, Hübner and Gössling, 2012, Klint et al., 2012 and Uyarra et al., 2005) and tourism in mountain regions (e.g. Prideaux et al., 2010 and Scott, Jones, et al., 2008).

2.2. Climate change perception studies in snow-based destinations

In snow-based destinations, demand-side climate change perception studies have been conducted in Australia (e.g. König, 1998 and Pickering et al., 2010), New Zealand (e.g. Prince, 2010), North America (e.g. Dawson, Havitz, & Scott, 2011) and Europe (e.g. Behringer et al., 2000 and Unbehaun et al., 2008). In Australian ski resorts, there has been a growing amount of supply-side climate change perception studies conducted among tourism managers (e.g. Bicknell and McManus, 2006, König, 1999, Morrison and Pickering, 2013, Roman et al., 2010 and Turton et al., 2010). An overview of European supply-side perception studies conducted among various stakeholder groups in the past decade is given in Table 1.

Table 1. Study designs of previous perception surveys among key players in snow-based winter tourism.

Author(s) (year)	Study area	Sample	Research method	Interview periods
Bürki (2000)	(Central) Switzerland	Tourism representatives $N = 17$	Focus groups	Summer 1998
Saarinen & Tervo (2006)	(Northern) Finland	Nature-based tourism operators $N = 19$ (9 operating on snow and winter activities)	In-depth interviews; semi-structured questionnaire	Spring & early summer 2005
Abegg et al. (2008)	Switzerland	Ski lift operators $N = 160$	Questionnaire	Summer 2007
Wolfsegger et al. (2008)	Austria	Ski resort managers $N = 36$	Online questionnaire	September/October 2005
Luthe (2009)	France, Italy, Austria, Switzerland	Ski area managers $N = 20$	In-depth interviews	May to September 2007
Saarinen & Tervo (2010)	Finland	Winter tourism businesses $N = 173$	Questionnaire, thematic interviews	Summer 2007
Strobl et al. (2011)	Austria	Destination stakeholders $N = 65$ (in three destinations)	Moderated discussions in three workshops	March 2010
Brouder & Lundmark (2011)	(Northern) Sweden	Winter-oriented tourism businesses $N = 63$	Online questionnaire	December 2008
Tervo-Kankare (2011)	Lapland, central Finland	Tourism stakeholders: businesses, development officials, related industry $N = 19$ (in two destinations)	Semi-structured face-to-face interviews	Spring & fall 2009, spring 2010

Bürki (2000) used focus groups to investigate how tourism representatives in Central Switzerland perceive the issue. Even at that time – long before climate change became the subject of today's broad media coverage – the author reported that "climate change has been recognised as a problem for [snow-based] winter tourism" (Elsasser & Bürki, 2002, p. 255). The survey participants stated that climate change was already affecting destination development. The group discussions also revealed the ambivalent attitude of tourism representatives toward climate change. On the one hand, potential consequences were downplayed. On the other, climate change was used to legitimate both the expansion of snowmaking facilities and the extension of ski areas to higher elevations. The tourism representatives were unanimous in their belief that Alpine winter sports can only survive if snow reliability is guaranteed.

This view is supported by Abegg, Kolb, Sprengel, and Hoffmann (2008). In a representative study among Swiss ski lift operators, the authors demonstrated that ski area managers trust the adequacy of snowmaking as a means of adapting to climate variability and change. Abegg et al. (2008) also reported that once prevailing defensive attitudes have given way to some kind of openness to the issue of climate change. They concluded that this openness and a high awareness of the problem in general are prerequisites for meeting the challenges that climate change is likely to pose for the snow-based winter tourism industry in upcoming decades. The results clearly indicated that Swiss ski lift operators take climate change seriously, with certain adaptation strategies already implemented and others planned.

In Austria, Wolfsegger, Gössling, and Scott (2008) conducted an online survey among managers of low-elevation ski areas. The results showed that climate change is not perceived as a substantial threat. The ski area managers were highly optimistic about the effectiveness of technical adaptation means, such as snowmaking, and they considered these to be sufficient to cope with climate change in the decades ahead. Sharing the costs of snowmaking with the accommodation industry was regarded as the most appropriate business strategy, followed by joining ski conglomerates and diversifying winter offerings.

A recent study by Strobl, Steiger, Peters, and Weiermair (2011) in three Austrian destinations employed moderated discussions about climate change and its impact on tourism. The authors found that climate change is a topic of discussion in these destinations, but mainly from a purely economic perspective focusing on how climate change influences the rates of return on investments in the long term.

Luthe (2009) conducted a transnational study of ski areas in France, Italy, Austria and Switzerland. Although the extent of change was unclear for many interviewees, the occurrence of climate change was not doubted. The ski area managers viewed high elevation and the possibility to make man-made snow to be crucial factors in meeting the challenges of climate change. They worried most about the increasing costs of technical adaptation and temperatures too high for snowmaking, but also about the indirect effects that climate change has on demand patterns as a result of snow-poor winters.

Saarinen and Tervo (2006) identified perception and adaptation strategies of nature-based tourism businesses with regard to climate change in Finland. The authors reported that the interviewees were generally aware of the issue, but at the same time highly sceptical. Half of the survey participants did not believe that the climate is actually changing and were doubtful about any significant effects on future tourism in their region. Climate change was seen as a minor threat, if a threat at all. The authors recognised a lack of adaptation planning and showed that adaptation strategies like snowmaking were undertaken to meet the challenges presented by general weather variations and market changes. Since many respondents were already used to adapting tourism operations to changing weather conditions in the short term (see also Tervo, 2008), they felt optimistic about being able to manage a gradual change in climate, if necessary.

Two years after this interview-based study, the authors conducted another survey among winter tourism businesses in Finland that revealed a growing awareness to changing climate (Saarinen & Tervo, 2010). Perceptions had changed significantly in this short time, with the majority of businesspersons now believing that the climate is changing. Asked about signals in their regions, they reported shorter winters and delayed arrival of permanent snow. More than half of the respondents felt that future conditions for nature-based tourism would be unfavourable. Discussing this change in perceptions, the authors critically mention the differences in study design and admit that a direct comparison of study results is not possible. However, they argue that the respondents' awareness of the issue may have been heightened by two relatively warm and snow-poor winters in 2005/2006 and

2006/2007 as well as by the growing public debate on climate change and increasing media coverage in Finland.

Brouder and Lundmark (2011) conducted a perception analysis in northern Sweden. In a quantitative survey of winter-oriented tourism businesses, they found that three-quarters of all respondents believe that some form of climate change is under way. The authors emphasised the importance of an intra-regional scale of analysis and demonstrated differences between perceptions in the inland areas and the coastland. However, the study subjects did not feel that climate change would have a drastic impact on the tourism sector in the coming decade. With regard to previous and future strategies, the authors stated that tourism businesses "intend to remain as long as possible and will adapt during the season depending on conditions with no strong evidence of strategic, long-term planning" (Brouder & Lundmark, 2011, p. 929).

The most recent investigation of the issue was conducted by Tervo-Kankare (2011), who interviewed tourism stakeholders in two Finnish destinations. The author reported that concern about potential impacts of climate change is mainly induced by extraordinary weather events and belated starts to the Christmas season. The level of climate change-related activity in the destinations was found to be low and largely dependent on individual ambition. Although tourism stakeholders generally had a good understanding of the issue, climate change was considered "to be a minor factor among all factors affecting the destinations' future and development plans" (Tervo-Kankare, 2011, p. 407). One main conclusion of the survey was that tourism-related climate change awareness has not yet found its way to the practical level.

The different research designs and methodologies, as well as the diverging characteristics of the tourism system itself in the respective study areas, hamper a detailed comparison of survey results. Moreover, there are indications that differences in perceptions may also be influenced by the winter weather preceding the investigation. The studies reporting a relatively high awareness of climate change were conducted after the analogue winter seasons of 1997/1998 (Bürki, 2000) and 2006/2007 (Abegg et al., 2008, Luthe, 2009 and Saarinen and Tervo, 2010). An analogue winter season is an anomalously warm season that is likely to represent average future climate conditions (Steiger, 2011).

Like Abegg et al., 2008 and Luthe, 2009 pointed out that his survey was conducted in summer 2007 following the warmest winter season on record in many parts of the European Alps. This may have influenced the study results. In their comparative paper discussing the results of their perception studies conducted in 2005 and again in 2007, Saarinen and Tervo (2010) concluded that the two relatively warm and snow-poor winters in 2005–2006 may have contributed to increased risk awareness among study participants.

The literature reveals most steps taken to adapt to climate change to be reactive and highly individualistic for various reasons, such as the competitive nature of the ski industry with its short planning cycles or differing business models, geographic characteristics or government jurisdiction (Abegg et al., 2007, Scott and McBoyle, 2007, Tervo-Kankare, 2011 and Trawöger, 2011). As Scott & McBoyle (2007) state, it is "difficult to sort out adaptations in the ski industry that were driven mainly by climate and those that are the result of other non-climate factors" (p. 1426). Most strategies currently aim at some form of risk reduction in relation to climate variability.

The present study was conducted after an average winter with regard to snowfall and temperature. Therefore, the participants' perceptions were not affected by having experienced an exceptionally snow-poor and warm winter season immediately prior to the study. The preceding winter of 2009/2010 was characterised by a warm and dry November. In most ski resorts, a very cold December had already made snowmaking possible at the beginning of the season. Persistently good snow conditions until Easter in late March produced good results for Tyrolean winter tourism.

3. STUDY AREA

The study area, Tyrol, is a state situated in the western part of Austria (Fig. 1). Tyrol holds a dominant position in Alpine tourism in general and snow-based winter tourism in particular. In 2010, the region had a total of 43 million overnight stays, which accounted for as many as one-eighth of all overnight stays estimated in the Alps (Statistics Austria, 2011). Tyrol ranks seventh among the top 20 tourism regions in the EU-27 with regard to nights spent by non-residents in hotels and campsites (Eurostat, 2011). On a national level, Tyrol generates about 40% of Austria's tourism turnover and holds one-third of the country's beds.

The total number of 24 million domestic and international skier visits recorded in Tyrol (WKO, 2012) represents 47% of Austrian, 13% of Alpine and 6% of global skier visits.

Figure 1. Sketch map of the study area.

The economic significance of tourism in the state itself varies regionally. While tourism contributes an average of 16% to the Tyrolean GDP, considerably higher percentages are reported by various counties. In valleys such as the Ziller Valley, the Ötz Valley or the Paznaun, tourism destinations report that an average of 57% of the population is employed in the tourism industry, with some figures as high as 86% (Trawöger, 2010). These values are well above the average of 35% (ÖHV, 2008) reached in Austrian winter tourism communities (i.e., communities with more than 60% of overnight stays during the winter season).

With its location in the European Alps, the climate of the area includes oceanic, continental, polar and Mediterranean influences (Beniston, 2006). The Alpine climate in Tyrol is generally characterised by short and humid summers, with precipitation peaking in June and July. Due to orographic effects, precipitation exhibits wide discrepancies between various sites over very short distances (Beniston, 2006 and Fliri, 1975). Precipitation is highest along the northern rim of the Alps and lowest in inner-Alpine dry valleys (Fliri, 1975 and Frei and Schär, 1998) like the Inn Valley or the Ötz Valley. Temperatures generally show a low degree of spatial variation, with an average annual temperature of about 8–9 °C in lowlands. In the winter months, spatial differences are frequently caused by atmospheric inversion (lower temperatures in lowlands) and Foehn effects (higher temperatures). At higher altitudes, the cold periods are considerably longer, with permanent snow cover from November until April or even later.

Some recent investigations focus on the potential impacts of climate change on the ski industry in the study area (Steiger, 2010, Steiger, 2011, Steiger and Trawöger, 2011 and Trawöger, 2010). Steiger and Stötter (2013) applied a ski season and snowmaking simulation model to the Tyrolean ski areas. Two indicators were used to assess snow reliability: snow reliability according to the 100-days rule by Abegg et al. (2007) and ski operation maintained in the Christmas period in 75 per cent of all winters, according to Scott, Dawson, et al. (2008). Given that ski areas are equipped with snowmaking facilities, all ski areas in the study area are snow-reliable until the 2030s (high emission scenario) and 2040s (low emission scenario). At the end of the century, only between 28% (snow-reliable in the Christmas period) and 65% (snow-reliable according to the 100-days-rule) are snow-reliable in the low emission scenario. In the high emission scenario, the share of snow-reliable ski areas at the end of the century decreases to 4–14%. Snow production would need to be increased substantially because Steiger and Stötter (2013) assume a doubling of snow production in half of the ski areas by the 2050s and a tripling by the 2080s in the high emission scenario.

Based on the same ski season simulation model, Steiger & Trawöger (2011) developed a multi-level snow reliability classification scheme and related it to the economic significance of winter tourism in Tyrolean tourism board regions. The results indicate that some areas in the northwest (Tannheimertal) and northeast (Kitzbühel) of Tyrol will experience high vulnerability by the 2030s. Destinations in these areas are characterised by short- to medium-term snow reliability combined with a strong regional economic dependency on winter tourism.

4. STUDY DESIGN AND RESEARCH METHODS

4.1. Survey sample

The study design and research methodology are illustrated in Fig. 2. The survey sample was determined on the basis of a multivariate data analysis (Trawöger, 2010) that identified four groups of Tyrolean ski resorts differing in their projected vulnerability to climate change. The vulnerability defined by Trawöger (2010) results from the combination of economic significance of winter tourism and snow reliability derived from the ski season model SkiSim 2.0 by Steiger (2010).

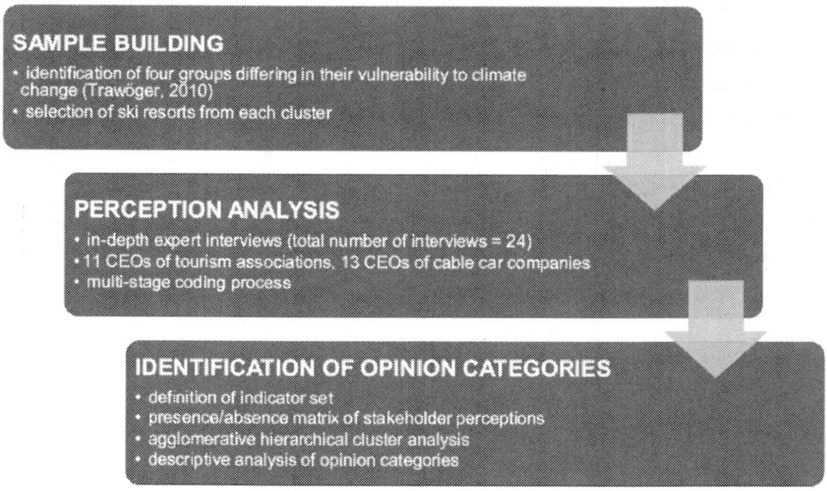

Figure 2. Study design and research methods.

Thus, the sample used for the study's perception analysis includes representatives of ski resorts most at risk from changing climatic conditions, as well as from areas facing challenges only in the long term. This type of survey sample ensures a mix of ski resorts that significantly differ from each other in many respects (i.e., regional importance of winter tourism, tourism intensity, elevation, size, snow reliability).

A total of 24 experts were willing to participate in the study, selected from ski resort clusters in Trawöger (2010). The survey includes 13 Chief Executive Officers (CEOs) of cable car companies and 11 CEOs of regional tourism associations. Cable car companies are often among the leading companies in a region. They play a central role in the development of the respective regions, especially in destinations focusing on ski tourism. Tyrolean tourism associations are public corporations that develop tourism strategies and conduct marketing activities in their regions. Hence, representatives of two of the most important stakeholder groups in Tyrolean winter tourism were interviewed to assess how the growing issue of climate change is currently being addressed. The hotel industry, gastronomy businesses and ski schools also represent important stakeholder groups, but their individual significance varies regionally.

4.2. Perception analysis

Using a qualitative approach, face-to-face in-depth expert interviews were conducted in German between March and August 2010. Guidelines, in the form of a topical pre-structure, ensured interview comparability. The open and semi-open questions were grouped into four topic areas: *destination positioning, market trends, perception of climate change* and *strategic planning*. The findings presented in this paper focus on the participants' responses to the latter two topics, including the following key questions:

- Do the stakeholders believe that some form of climate change is under way?
- Have they identified or do they expect signs of climate change in their regions?
- Have any climate change-related adaptation strategies been implemented or is their implementation at least anticipated?

Several pre-test interviews were conducted to exclude or identify initially problematic, overly complex or incomprehensible questions. The pre-tests showed that it was crucial to explain technical terms like *adaptation* in layman's terms to ensure that the participants actually understood the questions. Moreover, the feedback, especially to the semi-open questions, was very important to designing the final interview guidelines. Here, the pre-test participants critically commented on missing or ambiguous items. Another important insight from the pre-test interviews relates to the manner in which the interviewees were addressed. Focusing on the viability of future ski tourism in a broad sense, rather than just discussing the controversial and often negatively connoted topic of climate change, served as an ice-breaker (see also *6. Discussion*). The interviewees were assured full anonymity, which in many cases was a prerequisite for their willingness to participate. Consequently, the study results contain no direct references to individual regions, destinations or stakeholder identities. The sample building process ensured the inclusion of ski areas from throughout the study area, including about a third of all Tyrolean tourism associations ($N = 11$ out of 34).

The applied analysis procedure comprised a multi-stage coding process devised by Meuser and Nagel (1991) and included the following stages: *transcription*, *paraphrasing*, *coding* and *topical comparison*. The text analysis software MAXQDA was used for the evaluation.

Table 2. Indicators of stakeholder perception of climate change.

Number of indicator	Description
Indicator 1	I have already been involved with climate change in my everyday work.
Indicator 2	I feel well informed about climate change.
Indicator 3	I trust climate projections and impact studies dealing with climate change and winter tourism.
Indicator 4	I believe there is a trend toward (global) warming.
Indicator 5	I can identify signs of climate change in my region/ski area (apart from receding glaciers and melting permafrost).
Indicator 6	I expect signs of climate change in my region/ski area in the future.
Indicator 7	I am convinced that climate change poses a real and immediate threat to ski tourism in its present form.
Indicator 8	Climate change is a topic of discussion in my region/ski area.
Indicator 9	Coping with climate change already is or will soon be part of our risk management.
Indicator 10	Snowmaking is an appropriate means of adapting to climate change.

4.3. Identification of opinion categories

To identify different categories of stakeholders with common sets of characteristics, a cluster analysis was performed. This type of explorative multivariate analysis aims to divide objects into groups using a designated list of variables (Everitt et al., 2001 and Kaufman and Rousseeuw, 1990).

For this purpose, a set of 10 indicators was defined that represented key attitudes of stakeholder perception of climate change (Table 2). These key attitudes were carefully identified from the respective literature (see *1. Introduction*) by analysing the aspects in which perceptions of climate change and current ways of dealing with the issue among stakeholders differ from each other. Thus, the indicators allow a differentiated assessment of perceptions of climate change.

Using the coded interview transcripts, these indicators were determined to be present (scored 1) or absent (scored 0), which resulted in the formation of a binary presence/absence matrix (Table 3). This matrix served as input to the subsequent cluster analysis.

Table 3. Presence/absence matrix of stakeholder perceptions.

Stakeholder	Indicator 1	Indicator 2	Indicator 3	...	Indicator 10
A	1	0	1		0
B	1	1	1		0
C	1	0	0		0
...					

The term *cluster analysis* comprises different group-forming techniques that can be distinguished mainly by applied proximity measures and fusion algorithms (Everitt and Dunn, 2001, Hair et al., 2010 and Härdle and Simar, 2003). In this study, the cluster analysis was performed using the *simple matching coefficient* as a proximity measure and the *group average linkage* fusion algorithm.

The simple matching coefficient is one of the most commonly used similarity coefficients for binary data formats and is usually employed when co-absences of attributes provide information about the similarity of two objects (Everitt et al., 2001, Johnson and Wichern, 1992 and Kaufman and Rousseeuw, 1990). This means that both co-presences (indicator score = 1 for two stakeholders) and co-absences (indicator score = 0 for two stakeholders) provide equal evidence of similarity. Other similarity coefficients discount 0–0 matches or even disregard them completely (Johnson & Wichern, 1992), which would have been counter-productive in the present analysis.

The group average linkage fusion algorithm is an agglomerative hierarchical fusion method. Agglomerative algorithms start a merging process from each single object as a separate cluster (bottom-up) and identify successive clusters by means of formerly established clusters. Thus, objects of already existing clusters are not separated during the further clustering process (Everitt et al., 2001 and Johnson and Wichern, 1992). The group average linkage algorithm is also known as the *unweighted pair-group method.* It uses an average approach that computes all distances between pairs of objects in any two clusters and fuses the nearest two clusters. It tends to join clusters with small variances and is considered to be relatively robust (Everitt et al., 2001).

The cluster analysis was conducted using PASW Statistics 18 software. The final number of clusters was determined by analysing the resulting dendrogram that illustrates the fusion process. The four-cluster solution was found to provide a reasonable summary of the data, corresponding to a 70% similarity level of fused groups. This means that combined clusters show at least 70% similarity, as is the case until stage 20 of the fusion process.

In the last step, the four clusters were labelled to reflect the nature of cluster members. The attributed adjective describes the members' general attitude towards climate change (*convinced, annoyed, ambivalent*), whereas the noun refers to their current way of dealing with the issue (*planners, deniers, optimists, wait-and-seers*). To illustrate the cluster characteristics, some remarks made by the participants are presented in the cluster description.

5. RESULTS

5.1. General perceptions

All CEOs have already encountered the topic of climate change in their everyday work. They are mainly involved with calming the debate on the issue in communications with the banking industry and – in snow-poor winters – also with their markets and the media. In the context of capital budgeting, investments and ecological aspects (e.g., water quality and availability), climate change persistently plays only a tangential role.

The vast majority of the interviewees (22 out of 24) stated that they do not trust studies on climate change and snow-based winter tourism.

Analysis of the interviews reveals that this is primarily due to feelings of confusion caused by conflicting expert opinions depicted in the media and problems in differentiating between weather and climate, as the following statement shows:

"We are confronted with conflicting opinions. A few years ago they said we would have less snow. We had a lot of snow and they told us it was an exception. We had five exceptions in a row. They also said it would get warmer and then we had a freezing cold winter. That doesn't increase our confidence in studies. We can't estimate the long-term effects of climate change from the results we have received so far."

CEO, cable car company

Despite their scepticism toward the results of studies on climate change and the ski tourism industry, the CEOs indeed believe that some form of climate change is under way. Of the study participants, 20 out of 24 believe there is a trend toward (global) warming, although only 10 can already identify clear signs of climate change in their regions or ski areas, other than receding glaciers and melting permafrost. However, a majority of 14 interviewees anticipate signs of climate change in their region or ski area in future. Most CEOs expect a likely increase in extreme events but do not believe in highly negative consequences for ski resort operations.

Climate change is not perceived as a real and immediate threat to ski tourism in its present form. Coping with the issue is currently not an integral part of risk management nor will it be in the next few years for 19 of the interviewees. Consequently, tangible adaptation efforts have as yet not been discussed in most of the areas. Only one adaptation strategy is mentioned by some CEOs, namely the improvement of snow production techniques. There is unanimous agreement on the importance of snowmaking when operating a ski resort, even though not all managers believe in its adequacy as a means of adapting to climate change, as will be shown in the following section.

5.2. Opinion categories

Four categories of CEOs with shared characteristics were identified: *convinced planners*, *annoyed deniers*, *ambivalent optimists*, *convinced wait-and-seers*. An overview of cluster characteristics is given in Table 4 by listing the modes of the respective indicators (columns *I 1, I 2, ... I 10*) and the stakeholder group (column *Group*).

Table 4. Cluster characteristics (used measure of central tendency: mode; CC...cable car company CEOs; TA...tourism association CEOs) (I... Indicator (see Table 2); 0 = indicator is absent; 1 = indicator is present).

Cluster	I 1	I 2	I 3	I 4	I 5	I 6	I 7	I 8	I 9	I 10	Group
A	1	1	0	1	1	1	0	1	1	1	CC
B	1	0	0	0	0	0	0	0	0	0	[a]
C	1	1	0	1	0	0	0	1	0	0	CC
D	1	0	0	1	0	1	0	0	0	0	TA
Total	1	0	0	1	0	1	0	1	0	0	[b]

[a] Cluster B consists of only three members: 1 cable car company, 2 tourism associations.

[b] 11 tourism associations, 13 cable car companies.

5.2.1. Cluster A – convinced planners (*N* = 5; 5 cable car companies, 0 tourism associations)

Cluster A is made up of cable car company CEOs only and its members appear well informed about climate change. All of them are convinced that (global) temperatures will rise. The CEOs observe and expect signs of climate change in their ski areas:

For my part, I believe that something is under way. [...] Apart from all these scientific investigations the most genuine signs – anyone can identify – are found in nature. Things like plants moving up in elevation, you once found only on the valley floor, or new species – animals and plants – at higher altitudes, where conditions are cooler. It's things like that I believe.

CEO, cable car company

Although – like all other interviewees – *convinced planners* do not consider climate change to be an immediate threat to ski tourism, they are convinced that changing climatic conditions will affect the ski industry in the mid- and long-term. They report that climate change is a topic of discussion and they believe that they will need to cope with it as part of their companies' risk management in the near future (for two out of five, this is already true). The managers in Cluster A feel prepared because they are constantly improving snow production techniques with the aim of maintaining services as they are. Snowmaking is considered an appropriate means of adapting to climate change. The risk that temperatures could be too high for snowmaking is met by a focus on what they call *snow management*. This term designates further improvements in snowmaking and slope-grooming techniques as well as attempts to reduce snow production time. The snowmaking facilities in a ski area should be able to complete base-layer snowmaking (i.e., the production of 30 cm of snow) in only 72 h. The CEOs in Cluster A are convinced that climatic conditions in the early part of the season before Christmas will in the future ensure adequate temperatures for snowmaking in this short period of three days:

We can control an increase of 1°C, even 1.5°C with snowmaking facilities without using chemical additives. Here, the snowmaking possibilities grow with the problem. If we consider an increase of 2°C then, if you pardon my saying so, the world will have other problems than whether skiing is still possible in the Alps. If temperatures increase more than 2°C, winter tourism becomes a luxury.

CEO, cable car company

They show themselves to be nearly independent of natural snow, as the following statement emphatically underlines:

In fact we don't need snow, we make snow. Too much natural snow is bad for our business, because it means higher costs for slope grooming. Skiers only complain about natural snow pistes. They want smooth slopes that we can only provide with the help of machine-made snow. It sounds absurd, but the best scenario for us is less natural snow, low temperatures for snow production and lots of sun.

CEO, cable car company

Asked about their estimations of the potential for all-year tourism or increases in summer season figures, the CEOs in Cluster A voiced their doubts that snow management can save the day if climate change should cause real problems for ski tourism:

As far as increases in summer season figures are concerned, there are limits. In summer we sell nature, silence and relaxation. Even if we manage to get 30,000 hikers or mountain bikers in the area, that's not manageable. For us summer can be a nice extra income, but it can't – in any way – compensate for big problems in winter. We have ideas on how to work around the problem, but definitely no idea how to solve a real problem for ski tourism.

CEO, cable car company

Mentioned forms of behavioural adaptation are cooperatives and mergers with other areas as well as improvements in media communication and service quality.

5.2.2. Cluster B – annoyed deniers (*N* = 3; 1 cable car company, 2 tourism associations)

Cluster B has only three members and is thus the smallest group. Like all other interviewees, the managers in this cluster have already encountered climate change, but only in the sense that they calm down the debate on the issue in communications with the media, the tourism markets and the banking sector:

We try to calm down the debate, to state that there have always been snow-poor winters and that they will happen again and again and show that we have taken precautions in the form of snow production. The worst thing was the banks: they set altitude limits for snow reliability in their risk assessments.

That was really dangerous.

CEO, cable car company

The two CEOs of tourism associations do not feel that their tourism organisations keep them well informed about climate change, while the CEO of the cable car company feels well informed. However, none of them believes that a (global) warming trend is under way or has yet recognised signs of climate change in their regions or ski areas. The *annoyed deniers* do not expect any negative consequences for ski tourism in the future and view climate change as mainly a media topic and accuse scientists of inciting panic.

For me climate change is a media catchphrase. Moreover, many would-be scientists are playing to the gallery and misusing the issue for their own benefit. I had discussions with some of them and couldn't believe my ears. We live in nature, we live off the snow and we are not blind to what is happening. And nothing is happening. Temperatures and snow conditions always follow an up-and-down pattern.

CEO, cable car company

Climate change is not discussed in their regions or ski areas and is also not an agenda for risk management, neither now nor expected to be in the future. The CEOs in Cluster B generally feel annoyed by climate change discussions and do not see any reason or need to implement any kind of adaptation or mitigation strategies.

5.2.3. Cluster C – ambivalent optimists (*N* = 5; 4 cable car companies, 1 tourism association)

The *ambivalent optimists* feel they are well informed about climate change and believe that a (global) warming trend is in progress, although they cannot observe any signs in their regions or ski areas so far. Moreover, they do not expect any relevant future changes to occur in their region due to climate change. They are staying calm and, although they can imagine unfavourable global consequences induced by climate change, they do not think that ski tourism in its present form will be adversely affected.

According to the *ambivalent optimists,* climate change is repeatedly a topic of discussion in their everyday work, but mainly in the field of communication with their markets in snow-poor winters. None of them sees climate change as a risk management priority. The persons in Cluster C agree with the *convinced planners* (Cluster A) on the importance of snowmaking, but do not perceive it as a means of coping with climate change. For them, large-scale snowmaking is a pre-requisite for a competitive ski destination and predominantly a matter of quality management:

Nobody wants to ski on the natural snow slopes of the 1970s. We have to produce technical snow anyway to meet our customers' demands. That's why snowmaking developed – long before there were any climate change discussions. We do large-scale base-layer snowmaking in autumn anyway, no matter how much natural snow the winter brings.

CEO, cable car company

Some of the managers in Cluster C regard their optimistic attitude as a necessity in their profession. Finding themselves in a competitive environment may force managers to show some form of calculated optimism, which hampers a profound discussion of climate change matters:

If we acknowledged problems caused by climate change, we would be pegged as prophets of doom who foul our own nest. So, even if we already faced problems, I doubt that these would be discussed offensively for the time being.

CEO, cable car company

5.2.4. Cluster D – convinced wait-and-seers (*N* = 11; 3 cable car companies, 8 tourism associations)

This largest cluster is dominated by CEOs of tourism associations (eight out of 11) who do not feel well informed about climate change. Along with the *ambivalent optimists* they are convinced a global warming trend is under way, but cannot identify any signs of climate change in their regions. However, in contrast to the persons in Cluster C, the *convinced wait-and-seers* expect future challenges induced by changing climatic conditions. Some of them even question the focus on skiing and snow in the Tyrolean winter tourism industry. They also express their doubts about the effectiveness of snowmaking as a means of coping with climate change:

It would be presumptuous to say that the ski areas have taken precautions against climate change, because we haven't, we are not prepared for it. [...] Snow production may be a main strategy, but it is a strategy to compensate for a lack of snow, not for coping with climate change. We have not yet taken any real precautionary steps against a marked warming trend.

CEO, tourism association

Despite their recognition that climate change is under way and indications of a critical attitude towards technical means of adapting, they deal with the topic on a personal level and pursue a wait-and-see strategy. This is clearly shown by the fact that climate change is not a topic of discussion in these areas and is also not considered a risk management priority in the short- or mid-term. They agree with the *convinced planners* on the present lack of viable alternatives to ski tourism. Only one person in Cluster D expresses his visions more specifically:

There are no alternatives to skiing so far. We need intelligent, innovative products and specialisation, but there is little progress so far. We [Tyrol] should become pioneers in environmental management, but we don't get enough political support with regard to climate protection. The stakeholders and politicians are too short-term focused.

CEO, tourism association

5.3. Differences between the stakeholder groups

The clusters differ significantly in relation to the predominant stakeholder group (Table 5). Cluster A (*convinced planners*) is made up of cable car company managers only. Cluster C (*ambivalent optimists*) is also dominated by cable car company CEOs (four out of five). Tourism association managers mainly play the role of *convinced wait-and-seers* in Cluster D, in which they account for eight of the 11 cluster members.

Table 5. Cluster affiliation of survey participants (CC...cable car company CEOs; TA...tourism association CEOs).

Cluster	CC ($N = 13$)	TA ($N = 11$)
A – Convinced planners	5	0
B – Annoyed deniers	1	2
C – Ambivalent optimists	4	1
D – Convinced wait-and-seers	3	8

In addition to the cluster differences already mentioned in section *5.2, Opinion Categories*, two aspects stand out. On the one hand, the managers of tourism associations tend to be more critical about technical adaptation strategies in general and snowmaking in particular than do CEOs of cable car companies. On the other hand, the CEOs of tourism associations do not feel well informed about climate change (10 out of 11), while cable car company CEOs generally do (10 out of 13).

6. DISCUSSION

The findings of various perception studies (see *2. Literature review*) indicate a growing awareness of climate change matters among stakeholders in the snow-based winter tourism industry. Swiss cable car companies have taken on a pioneering role in the European Alps as far as the implementation of climate change in corporate planning is concerned. Abegg et al. (2008) report that between 2002 and 2007 nearly half of all investigated Swiss cable car companies (49%) gave consideration to an ongoing warming trend in corporate planning, while a majority (79%) intended to do so in the upcoming years.

The present study documents differing views among Tyrolean stakeholders, with hardly any climate change adaptation strategies implemented or planned. Although a majority (83%) of the interviewees believe that a global warming trend is at least in progress, 79% state that climate change is not and will not be a risk management priority in upcoming years. This contradiction between a generally growing awareness of climate change and the non-existence of action has also been recently identified by Tervo-Kankare (2011) in two Finnish destinations.

But why do the identified risk perceptions deviate so strongly from scientific estimates of the challenge of climate change? Judging from the survey, there are various reasons for the current lack of interest and efforts to adapt to climate change. These range from very short planning cycles of less than 10 years and a still largely low degree of economic suffering in the ski industry to the general attitude of stakeholders, who feel obliged to believe in the future of ski tourism and to demonstrate optimism in a competitive environment (see also Trawöger, 2011).

However, another main reason for the lack of adaptation strategies is the fact that most stakeholders simply do not perceive climate change as a real threat, in many cases not even as a risk to their businesses or regions. What might at first appear to be contradictory to the generally growing awareness in the literature is explained by characterising the kind of awareness observed in the Tyrolean ski tourism industry. The stakeholders' awareness identified in this study seems to be mainly limited to the general issue of climate change in the sense of *Yes, I believe that something is under way, primarily on a global level*. Most interviewees are not aware of potential branch- or region-specific consequences of climate change.

Shifting perceptions and attitudes pose another difficulty in assessing adaptive capacities in the ski industry. This crucial matter immediately prompts the question as to what major factors influence climate change perception and awareness among stakeholders. Is it really scientific publications, such as IPCC reports, information disseminated by authorities and media coverage that determine climate change perception on the supply side or is it tangible signs of climate change that are identified in the regions?

As pointed out in *1. Introduction*, perceptions of climate change are also likely to be influenced by the character of the winters preceding the surveys, with anomalously warm and snow-poor winters triggering a greater awareness. This also applies in the reverse case, with relatively cold and snow-rich winters reducing awareness of climate change. The findings of the present survey support this hypothesis, with many interviewees stating that discussions about climate change have petered out in the last five years, corresponding with no anomalously warm or snow-poor winters.

The immediate experience of winter weather in the region proves to be one of the key drivers of climate change perception among the interviewees. Many participants refer to recently experienced weather characteristics in previous winter seasons when asked about their perception of climate change. Although some interviewees already report direct effects of climate change in their regions, including rising tree or snow lines, receding glaciers or melting permafrost, it is the personal experience of winter weather characteristics that increases or decreases their awareness of climate change. This is confirmed in a recent interview with the head of the tourism board in Kitzbühel, who reduces the issue of climate change to a perception problem on the demand side. He stresses that, despite short-term losses, the coldness of the preceding winter was a huge benefit from a tourism perspective. Moreover, he states that potential consequences of climate change had been on peoples' minds for years, including projections that there would be no more snow. He further mentions that this notion had completely disappeared and that people could be more than optimistic for the upcoming winter season. (Christler & Zierl, 2012).

These study results are consistent with the findings by Ratter, Philipp, and Storch (2012), who found a decline in attention to and concern about climate change in recent years in many parts of the world. The authors relate this development to various causes, including information fatigue by the media and influences of regional weather conditions. Their analysis shows that a long-term increase in scientific confidence and media coverage of the issue does not consequentially result in higher public concern. However, Ratter et al. (2012) note that the decline may be only intermittent. Recent studies by Krosnick, Holbrook, Lowe, and Visser (2006) and Li, Johnson, and Zaval (2011) also suggest a connection between local weather and concern about global warming.

The present study shows that knowledge about climate change as a global phenomenon does not automatically trigger action on a regional level. According to the interviewees, uncertainties and the lack of regional projections and local impact studies make it difficult for them to implement adaptation strategies in their planning. Assuming that anomalously warm and snow-poor winters are a determining factor, a series of such winters causing snowmaking problems is likely to heighten awareness among stakeholders. Saarinen and Tervo (2010) also assume a connection between the growing awareness of climate change within a short period of time and the immediate experience of two warm and snow-poor winters.

The literature on risk perception provides cognitive and cultural explanations for the failure to address climate change, which are also relevant in the context of this study. One cognitive bias, labelled unrealistic optimism, optimism bias or illusion of invulnerability, was first demonstrated by Weinstein (1980). He reported that people generally tend to be unrealistically positive about the future rather than objective (see also Bazerman, 2006, Berger et al., 2011 and Wagner and Zeckhauser, 2012). Weinstein (1980) ascribed individual unrealistic expectations about the future to errors in judgement, influenced by such factors as perceived probability and personal experience. According to Costa-Font et al. (2009), new risks like climate change seem to be especially influenced by an optimism bias.

Recent research in neuroscience dealing with unrealistic optimism found that this mechanism is tied to diminished coding of undesirable information in the human brain (Sharot, Korn, & Dolan, 2011). The authors suggest that positive illusions are the result of selective update failures. Consequently, people often underestimate future negative events even if they are confronted with information challenging their optimistic outlook. Cognitive dissonance is another closely related psychological phenomenon that helps to explain stakeholders' attitudes toward climate change: confronted with uncertainty, people tend to view challenges in a way that makes them most comfortable (Wagner & Zeckhauser, 2012).

Moreover, high degrees of perceived control lead to lower personal risk estimates and increased risk denial (Bazerman, 2006, Klein and Helweg-Larsen, 2002 and Sjöberg, 2000). Many stakeholders interviewed, especially the CEOs of cable car companies in Cluster A (*convinced planners*), think that they can control the risk of too high temperatures for producing man-made snow by further improvements in snowmaking and slope-grooming techniques. They also exhibit very optimistic views about ski tourism in a warmer climate. These findings are consistent with research on risk perceptions, which links optimism and perceived controllability over future events. For Gifford (2011), technosalvation – the belief that technology (nearly) alone can solve the problem – is one of the *dragons of inaction* limiting climate change mitigation and adaptation.

In the interviews, some other barriers to action and behavioural change were revealed that corroborate previous research. These include the lack of trust in scientific studies leading to risk denial (Gifford, 2008 and Sjöberg, 2000) and the human tendency to maintain the status quo, which is closely related to loss aversion (Bazerman, 2006 and Wagner and Zeckhauser, 2012).

The findings of this study also support previous research on public perceptions of climate change that identifies certain qualitative characteristics of the issue as sources of inaction. The majority of the CEOs could not make out clear signs of climate change that directly threaten operations in their areas and anticipate challenges mainly in the distant future. Several studies found this *delay effect* to be a major barrier to adapting to climate change (Etkin and Ho, 2007, Heath and Gifford, 2006, Leiserowitz, 2005, Leiserowitz, 2006, Sunstein, 2006 and Weber, 2006). People predominantly expect climate change to occur in the future and to affect geographically distant regions. Sunstein (2006) stresses that serious harms caused by climate change are not cognitively available to people, which prevents them from taking action.

For Bazerman (2006), the lack of vividness is one of the most fundamental sources of *predictable surprises* like climate change. The author uses this term to describe events that lead organisations and nations to react with surprise, despite people's general awareness of the challenge and strong scientific arguments for action. Weber (2006) argues that insufficient feedback from daily or yearly personal experience may prevent people from reacting to global warming – even those who are dependent on weather and climate like farmers or fishermen. These considerations also help to explain the inaction among Tyrolean winter tourism stakeholders. As long as effects of climate change are not resulting in direct consequences for tourism-related operations in the ski areas, they are unlikely to lead to climate change adaptation strategies.

This view is supported by Oppenheimer and Todorov (2006), who state that concerns are delayed until impacts are obvious and personal. Thus, Oppenheimer and Todorov (2006) stress the need to translate general concerns about climate change into local, personally relevant worries. Due to its complex, overarching and often abstract nature, the problem of climate change cannot be solved with traditional linear models (Etkin & Ho, 2007). It fits into what is labelled *post-normal science*, dealing with issues characterised by uncertainty, conflicting opinions and high decision stakes (Funtowicz & Ravetz, 1993).

The identified clusters in this study are quite homogenous with respect to the investigated stakeholder groups. While Clusters A and C are dominated by CEOs of cable car companies, Cluster D is mainly made up of CEOs of tourism associations. The interviews revealed that, apart from personal experience of recent and past weather conditions, the stakeholders tend to connect their attitudes to the ones predominating in their stakeholder group. This phenomenon of cultural polarisation is discussed by Kahan et al. (2012), who state that people consult others with whom they share close ties when acquiring scientific knowledge. Thus, they avoid dissonance and protect their standing within their groups. As Kahan (2012) argues, the individual is often confronted with partisan meanings of this kind: "if you are one of us, believe this; otherwise, we'll know you are one of them" (p. 255). In their investigation, Kahan et al. (2012) refer to the *tragedy of the risk-perception commons*, reflecting the conflict between the individual and collective levels of rationality.

Closely related phenomena, which help to explain different attitudes among groups or even nations, are those of group polarisation and confirmation bias. Sunstein (2006) states that group polarisation inevitably occurs in the context of risk perception. In group discussions, like-minded people tend to end up accepting a more extreme version of their predominant views. Hence, fear or neglect of certain events is amplified. This process interacts with what is known as confirmation bias – the human tendency to memorise information that confirms original beliefs (Klayman & Ha, 1987) – and social bandwagon or cascade effects (Sunstein, 2006).

Many stakeholders interviewed have a negative emotional connection to the issue of climate change in the sense that they associate it with media panic, threat scenarios or downright pessimism. Some even consider offensive discussions about climate change to tarnish their region's image and to befoul their own nest. An open-minded approach does not seem possible in many regions of the study area. Regarding the difficulties still surrounding the viability and validity of regional impact studies, new approaches on how to address stakeholder scepticism may pave the way for establishing adaptation strategies or at least generating discussion about climate change in the regions.

In this survey, the interviewees were addressed by focusing on the viability of future ski tourism in a broader sense, but also by discussing market trends, leadership in winter tourism destinations and matters of sustainability and risk management. These areas of interest served not only as an important source of information but also as ice-breakers for discussing the controversial and often negatively connoted topic of climate change. Such an approach may also be advantageous for studies in other areas characterised by hostile or indifferent attitudes towards climate change. Tervo-Kankare (2011) identified a similar recommendation expressed by the representatives of local development agencies in Lapland who regarded climate change "with its complex cause and effect relationship to be too broad, scary and paralysing an issue to handle, while pursuit of sustainability, for example, was easier for tourism entrepreneurs to understand" (p. 408). Many quotations in the literature (e.g. Bürki, 2000, Luthe, 2009, Saarinen and Tervo, 2010, Tervo-Kankare, 2011 and Trawöger, 2011) – especially those cited in surveys conducted after anomalously warm and snow-poor winters – show how emotionally charged the topic of climate change is.

7. CONCLUSION

This article examines the perceptions of climate change among winter tourism stakeholders in one of the top Alpine tourism regions. Tyrolean CEOs of cable car companies and tourism associations currently show themselves to be quite calm and relaxed in the face of an ongoing warming trend. The findings of the present survey indicate that climate change is not perceived as a threat. Those interviewed either feel prepared due to snowmaking (*convinced planners*) or do not feel that serious consequences for ski area operations are likely. The general belief that climate change is a global phenomenon has not yet developed into an awareness of potential branch- and region-specific consequences, nor has it prompted action on a regional level.

No adaptation strategies could be identified that were primarily adopted as a reaction to an already changing climate or in anticipation of future temperature or precipitation modifications. Large-scale snowmaking and improvements in slope grooming are predominantly undertaken to reduce dependency on natural snow and to enhance slope quality. Advances in snowmaking have already enabled many destinations to kick off the ski season with no natural snow at all, as the well-known Tyrolean winter resort Ischgl impressively showed in November 2011. It is no longer the lack of natural snow that prevents ski destinations from opening the season, but – if so – the inability to produce snow due to high temperatures. However, tourism association managers tend to take a more long-term perspective than cable car company managers and also show a more critical attitude toward technical adaptation strategies like snowmaking.

There is reason to assume that in the short and medium terms, coping strategies in the Tyrolean ski industry will remain highly individualistic and reactive. Recent cold or snowy winters proved to be a reason for climate change discussions to peter out in Tyrolean winter tourism.

It is clear from the above discussion that further study is needed in several fields of research: (a) regional scenarios of climate change impacts on winter and ski tourism; (b) investigations of the most influential drivers of climate change perception and ways to transfer scientific knowledge to practitioners; and (c) representative perception studies in various Alpine regions employing similar study designs in order to allow comparative assessment of adaptive capacities in different regions. With regard to perception analyses, particular attention should be given to the sampling process. The importance of certain stakeholder groups varies from region to region, and thus the selection of participants needs to follow the nature of the tourism systems in the investigated regions while considering the organisational structures and decision making processes in those areas.

There is evidence that particular attention should also be given to new forms of management that allow people to tackle complex problems like climate change successfully. Many current management models seem to have reached their limits. Being frustrated with management-as-usual, a group of scholars and business leaders recently defined "management moon shots" (Hamel, 2009, p. 1). They address a number of needs for behavioural change in management, which are also revealed in this study about ski tourism stakeholders. These include the priorities of decreasing fear and increasing trust to facilitate innovation and change, instead of mindlessly reinforcing the status quo, and to stretch executive time frames and perspectives.

The scientific task for and challenge of the upcoming years is to connect the general awareness of climate change as a global phenomenon to the already ongoing and potential branch- and region-specific effects of modified temperature and precipitation patterns. The results of the present survey suggest that only an increased knowledge of branch- and region-specific consequences of climate change related to snowmaking costs, water demand and the like will trigger a profound discussion of the issue and thus finally cause action to be taken.

ACKNOWLEDGEMENTS

The interviews were conducted as part of the project 'Sustainability of Ski Tourism in Tyrol and South Tyrol' funded by the Austrian Science Fund (FWF): L 348-G14. The author would like to thank all involved cable car companies and tourism region officials for their willingness to participate in the perception study, for thought-provoking discussions and their cooperation throughout the project. The author is also indebted to Johann Stoetter for supervising and supporting the project in a number of ways.

REFERENCES

1. Abegg, B., Agrawala, S., Crick, F., & de Montfalcon, A. (2007). Climate change impacts and adaptation in winter tourism. In S. Agrawala (Ed.), Climate change in the European Alps (pp. 25e60). Paris: OECD Publishing.
2. Abegg, B., Kolb, M., Sprengel, D., & Hoffmann, V. H. (2008). Klimawandel aus Sicht der Schweizer Seilbahnunternehmer. In T. Bieger, C. Laesser, & R. Maggi (Eds.), Jahrbuch der Schweizerischen Tourismuswirtschaft 2008 (pp. 73e83) (St.Gallen).
3. Baron, J. (2006). Thinking about global warming. Climatic Change, 77(1e2),137e150.
4. Bazerman, M. H. (2006). Climate change as a predictable surprise. Climatic Change,77(1e2), 179e193.

5. Becken, S. (2005). Harmonising climate change adaptation and mitigation: the case of tourist resorts in Fiji. Global Environmental Change, 15(4), 381e393.
6. Becken, S., & Hay, J. E. (2007). Tourism and climate change. Risks and opportunities. Climate change, economies and society. Clevedon: Channel View Publications.
7. Behringer, J., Buerki, R., & Fuhrer, J. (2000). Participatory integrated assessment of adaptation to climate change in alpine tourism and mountain agriculture. Integrated Assessment, 1(4), 331e338.
8. Belle, N., & Bramwell, B. (2005). Climate change and small island tourism: policy maker and industry perspectives in Barbados. Journal of Travel Research, 44(1), 32e41.
9. Beniston, M. (2006). Mountain weather and climate: a general overview and a focus on climatic change in the Alps. Hydrobiologia, 562(1), 3e16.
10. Berger, A., Brown, C., Kousky, C., & Zeckhauser, R. (2011). The challenge of degraded environments: how common biases impair effective policy. Risk Analysis, 31(9), 1423e1433.
11. Bicknell, S., & McManus, P. (2006). The canary in the coalmine: Australian ski resorts and their response to climate change. Geographical Research, 44(4), 386e400.
12. Braun, O. L., Lohmann, M., Maksimovic, O., Meyer, M., Merkovic, A.,Messerschmidt, E., et al. (1999). Potential impact of climate change effects onpreferences for tourism destinations. A psychological pilot study. Climate Research, 11, 247e254.
13. Brouder, P., & Lundmark, L. (2011). Climate change in Northern Sweden: intraregional perceptions of vulnerability among winter-oriented tourism businesses. Journal of Sustainable Tourism, 19(8), 919e933.
14. Buckley, R. (2008). Misperceptions of climate change damage coastal tourism: case study of Byron Bay, Australia. Tourism Review International, 12(1), 71e88.

15. Bürki, R. (2000). Klimaänderung und Anpassungsprozesse im Wintertourismus. In Ostschweizerische Geographische Gesellschaft (Ed.), Neue Folge (Vol. 6) (St.Gallen).
16. Buzinde, C. N., Manuel-Navarrete, D., Yoo, E. E., & Morais, D. (2010). Tourists' perceptions in a climate of change: eroding destinations. Annals of Tourism Research, 37(2), 333e354.
17. Christler, M., & Zierl, K. (2012, March 4). Bilanz nach Frost-Winter: Wer profitiert, wer verliert. Tiroler Tageszeitung. Retrieved from http://www.tt.com.
18. Costa-Font, J., Mossialos, E., & Rudisill, C. (2009). Optimism and the perceptions of new risks. Journal of Risk Research, 12(1), 27e41.
19. Dawson, J., Havitz, M., & Scott, D. (2011). Behavioral adaptation of alpine skiers to climate change: examining activity involvement and place loyalty. Journal of Travel & Tourism Marketing, 28(4), 388e404.
20. Dawson, J., & Scott, D. (2013). Managing for climate change in the alpine ski sector. Tourism Management, 35, 244e254.
21. Douglas, M., & Wildavsky, A. B. (1982). Risk and culture: An essay on the selection of technical and environmental dangers (1st ed.). Berkeley: University of California Press.
22. Elsasser, H., & Bürki, R. (2002). Climate change as a threat to tourism in the Alps. Climate Research, 20, 253e257.
23. Etkin, D., & Ho, E. (2007). Climate change: perceptions and discourses of risk. Journal of Risk Research, 10(5), 623e641.
24. European Commission. (2012). Supporting European tourism. Retrieved from http://ec.europa.eu/enterprise/sectors/tourism/index_en.htm.
25. Eurostat. (2011). Eurostat regional yearbook 2011. Luxembourg: Publications Office of the European Union.
26. Everitt, B. S., & Dunn, G. (2001). Applied multivariate data analysis (2nd ed.). London: Hodder Arnold.
27. Everitt, B. S., Landau, S., & Leese, M. (2001). Cluster analysis. London: Arnold.

28. Fischhoff, B., Slovic, P., Lichtenstein, S., Read, S., & Combs, B. (1978). How safe is safe enough? A psychometric study of attitudes towards technological risks and benefits. Policy Sciences, 9(2), 127e152.
29. Fliri, F. (1975). Das Klima der Alpen im Raume von Tirol. In Monographien zur Landeskunde (Vol. 1). Innsbruck: Universitätsverlag Wagner.
30. Frei, C., & Schär, C. (1998). A precipitation climatology of the Alps from highresolution rain-gauge observations. International Journal of Climatology, 18(8), 873e900.
31. Funtowicz, S. O., & Ravetz, J. R. (1993). Science for the post-normal age. Futures, 25(7), 739e755.
32. Gifford, R. (2008). Psychology's essential role in alleviating the impacts of climate change. Canadian Psychology/Psychologie Canadienne, 49(4), 273e280.
33. Gifford, R. (2011). The dragons of inaction: psychological barriers that limit climate change mitigation and adaptation. American Psychologist, 66(4), 290e302.
34. Gössling, S., Bredberg, M., Randow, A., Sandström, E., & Svensson, P. (2006). Tourist perceptions of climate change: a study of international tourists in Zanzibar.Current Issues in Tourism, 9(4), 419e435.
35. Hair, J., Black,W., Babin, B., & Anderson, R. (2010). Multivariate data analysis: A global perspective (7th ed.). Upper Saddle River, NJ: Pearson.
36. Hall, A. J., & Clayton, A. (2009). How will climate change impact on the tourism industry? Interviews with leading representatives of the travel and tourism industry in Jamaica. Worldwide Hospitality and Tourism Themes, 1(3), 269e273.
37. Hall, C. M. (2006). New Zealand tourism entrepreneur attitudes and behaviours with respect to climate change adaptation and mitigation. International Journal of Innovation and Sustainable Development, 1(3), 229e237.

38. Hall, C. M. (2011). Will climate change kill Santa Claus? The potential impacts of climate change on place competition. In M. Nuttall, H. Strauss, & K. Tervo-
39. Kankare (Eds.), Society, environment and place in northern regions (pp. 99e118). Oulu: University of Oulu, Thule Institute.
40. Hall, C. M., & Higham, J. (Eds.). (2005). Tourism, recreation and climate change. Clevedon: Channel View Publications.
41. Hamel, G. (2009). Moon shots for management: what great challenges must we tackle to reinvent management and make it more relevant to a volatile world? Harvard Business Review, 87, 91e98.
42. Härdle, W., & Simar, L. (2003). Applied multivariate statistical analysis. Berlin, Heidelberg, New York: Springer.
43. Heath, Y., & Gifford, R. (2006). Free-market ideology and environmental degradation: the case of belief in global climate change. Environment and Behavior, 38(1), 48e71.
44. Houghton, J. (2009). Global warming: The complete briefing (4th ed.). Cambridge: Cambridge University Press.
45. Hübner, A., & Gössling, S. (2012). Tourist perceptions of extreme weather events in Martinique. Journal of Destination Marketing & Management, 1(1e 2), 47e55.
46. IPCC (Intergovernmental Panel on Climate Change). (2000). Emissions scenarios. Cambridge: Cambridge University Press.
47. IPCC (Intergovernmental Panel on Climate Change). (2007). Contribution of working group II to the fourth assessment report of the intergovernmental panel on climate Change. Cambridge, United Kingdom and New York, NY, USA: Cambridge University Press. Retrieved from http://www.ipcc.ch.
48. Johnson, R. A., & Wichern, D. W. (1992). Applied multivariate statistical analysis. London: Prentice-Hall International.
49. Kahan, D. (2012). Why we are poles apart on climate change. Nature, 488(7411), 255.
50. Kahan, D., Peters, E., Wittlin, M., Slovic, P., Ouellette, L. L., Braman, D., et al. (2012). The polarizing impact of science

literacy and numeracy on perceived climate change risks. Nature Climate Change, 2(10), 732e735.

51. Kasperson, R. E., Renn, O., Slovic, P., Brown, H. S., Emel, J., Goble, R., et al. (1988). The social amplification of risk: a conceptual framework. Risk Analysis, 8(2),177e187.
52. Kaufman, L., & Rousseeuw, P. J. (1990). Finding groups in data: An introduction to cluster analysis. New York: Wiley.
53. Klayman, J., & Ha, Y. (1987). Confirmation, disconfirmation, and information in hypothesis testing. Psychological Review, 94(2), 211e228.
54. Klein, C. T., & Helweg-Larsen, M. (2002). Perceived control and the optimistic bias: a meta-analytic review. Psychology & Health, 17(4), 437e446.
55. Klint, L. M., Jiang, M., Law, A., Delacy, T., Filep, S., Calgaro, E., et al. (2012). Dive tourism in Luganville, Vanuatu: shocks, stressors, and vulnerability to climate change. Tourism in Marine Environments, 8(1), 91e109.
56. König, U. (1998). Climate change and tourism: investigation into the decisionmaking process of skiers in Australian ski fields. Pacific Tourism Review, 2(1), 83e90.
57. König, U. (1999). Climate change and snow tourism in Australia. Geographica Helvetica, 54(3), 147e157.
58. Kotlarski, S., Bosshard, T., Fischer, E., Lüthi, D., & Schär, C. (2010). Climate scenarios for the European Alps sand extreme events. Presentation at Swiss climate scenarios workshop, Zurich. Retrieved from http://www.c2sm.ethz.ch/news/scen_workshop/presentations/c2sm_ws10_kotlarski.pdf.
59. Krosnick, J. A., Holbrook, A. L., Lowe, L., & Visser, P. S. (2006). The origins and consequences of democratic citizens' policy agendas: a study of popular concern about global warming. Climatic Change, 77(1e2), 7e43.
60. Leiserowitz, A. (2005). American risk perceptions: is climate change dangerous? Risk Analysis, 25(6), 1433e1442.

61. Leiserowitz, A. (2006). Climate change risk perception and policy preferences: the role of affect, imagery, and values. Climatic Change, 77(1e2), 45e72.
62. Li, Y., Johnson, E. J., & Zaval, L. (2011). Local warming: daily temperature change influences belief in global warming. Psychological Science, 22(4), 454e459.
63. Lorenzoni, I., & Pidgeon, N. F. (2006). Public views on climate change: European and USA perspectives. Climatic Change, 77(1e2), 73e95.
64. Luthe, T. (2009). Vulnerability to global change and SkiSustain (Doctoral Dissertation). Cologne: Institute of Outdoor Sports and Environmental Science, German Sports University Cologne.
65. McEvoy, D., Cavan, G., Handley, J., McMorrow, J., & Lindley, S. (2008). Changes to climate and visitor behaviour: Implications for vulnerable landscapes in the North West Region of England. Journal of Sustainable Tourism, 16(1), 101e121.
66. Meuser, M., & Nagel, U. (1991). Experteninterviews e vielfach erprobt, wenig bedacht: Ein Beitrag zur Methodendiskussion. In D. Garz, & K. Kraimer (Eds.), Qualitativ-empirische Sozialforschung. Konzepte, Methoden, Analysen (pp. 441e 471). Opladen: Westdeutscher Verlag.
67. Morrison, C., & Pickering, C. M. (2013). Perceptions of climate change impacts, adaptation and limits to adaption in the Australian Alps: the skitourism industry and key stakeholders. Journal of Sustainable Tourism, 21(2), 173e191.
68. ÖHV (Österreichische Hoteliervereinigung). (2008). Österreichs Destinationen im Wettbewerb: Destinationsstudie und -karte der Österreichischen Hoteliervereinigung (Vienna).
69. Oppenheimer, M., & Todorov, A. (2006). Global warming: the psychology of long term risk. Climatic Change, 77(1e2), 1e6.
70. Pickering, C. M., Castley, J. G., & Burtt, M. (2010). Skiing less often in a warmer world: attitudes of tourists to climate change in an Australian ski resort. Geographical Research, 48(2), 137e147.

71. Prideaux, B., Coghlan, A., & McNamara, K. (2010). Assessing tourists' perceptions of climate change on mountain landscapes. Tourism Recreation Research, 35(2), 187e200.

72. Prince, B. W. (2010). Climate change adaptation and mitigation in New Zealand snow tourism. Industry Report, Ministry of Economic Development, Tourism Strategy Group. Wellington: School of Geography, Environment and Earth Sciences, Victoria University of Wellington.

73. Ratter, B. M., Philipp, K. H.,&Storch, H. (2012). Betweenhypeanddecline: recent trends

74. in public perception of climate change. Environmental Science & Policy, 18, 3e8.

75. Rauken, T., Kelman, I., Jacobsen, J. K. S., & Hovelsrud, G. K. (2010). Who can stop the rain? Perceptions of summer weather effects among small tourism businesses. Anatolia, 21(2), 289e304.

76. Roman, C. E., Lynch, A. H., & Dominey-Howes, D. (2010). Uncovering the essence of the climate change adaptation problemdA case study of the tourism sector at Alpine Shire, Victoria, Australia. Tourism and Hospitality Planning & Development, 7(3), 237e252.

77. Saarinen, J., & Tervo, K. (2006). Perceptions and adaptation strategies of the tourism industry to climate change: the case of Finnish nature-based tourism entrepreneurs. International Journal of Innovation and Sustainable Development, 1(3), 214e228.

78. Saarinen, J., & Tervo, K. (2010). Sustainability and emerging awareness to changing climate: tourism industry's knowledge and perceptions of the future of naturebased winter tourism in Finland. In C. M. Hall, & J. Saarinen (Eds.), Tourism and change in the polar regions: Climate, environment and experiences (pp. 147e164).

79. London: Routledge. Scott, D. (2011). Why sustainable tourism must address climate change. Journal of Sustainable Tourism, 19(1), 17e34.

80. Scott, D., Amelung, B., Becken, S., Ceron, J.-P., Dubois, G., Gössling, S., et al. (2008).
81. Climate change and tourism: Responding to global challenges. Madrid, Paris, Geneva: United Nations World Tourism Organization. Scott, D., Dawson, J., & Jones, B. (2008). Climate change vulnerability of the US
82. Northeast winter recreation-tourism sector. Mitig Adapt Strat Glob Change, 13(5e6), 577e596.
83. Scott, D., de Freitas, C., & Matzarakis, A. (2009). Adaptation in the tourism and recreation sector. In K. Ebi, & I. Burton (Eds.), Biometeorology for adaptation to climate variability and change (pp. 171e194). Dordrecht: Kluwer Academic Publishing.
84. Scott, D., Gössling, S., & Hall, C. M. (2012). Tourism and climate change: Impacts, adaptation and mitigation (1st ed.). London, New York: Routledge.
85. Scott, D., Jones, B., & Konopek, J. (2008). Exploring potential visitor response to climate-induced environmental changes in Canada's Rocky Mountain national parks. Tourism Review International, 12(1), 43e56.
86. Scott, D., & McBoyle, G. (2007). Climate change adaptation in the ski industry. Mitig Adapt Strat Glob Change, 12(8), 1411e1431.
87. Scott, D., McBoyle, G., & Minogue, A. (2007). Climate change and Quebec's ski industry. Global Environmental Change, 17(2), 181e190.
88. Sharot, T., Korn, C. W., & Dolan, R. J. (2011). How unrealistic optimism is maintained in the face of reality. Nature Neuroscience, 14(11), 1475e1479.
89. Sjöberg, L. (2000). Factors in risk perception. Risk Analysis, 20(1), 1e11.
90. Slovic, P. (1987). Perception of risk. Science, 236(4799), 280e285.
91. Slovic, P. (2000). The perception of risk. London, Sterling, VA: Earthscan Publications.

92. Smit, B., Burton, I., Klein, R., & Wandel, J. (2000). An anatomy of adaptation to climate change and variability. Climatic Change, 45, 223e251.

93. Smit, B., &Wandel, J. (2006). Adaptation, adaptive capacity and vulnerability. Global Environmental Change, 16(3), 282e292.

94. Statistics Austria. (2011). Arrivals, overnight stays. Retrieved from http://www. statistik.at.

95. Steiger, R. (2010). The impact of climate change on ski season length and snowmaking requirements. Climate Research, 43(3), 251e262.

96. Steiger, R. (2011). The impact of snow scarcity on ski tourism: an analysis of the record warm season 2006/2007 in Tyrol (Austria). Tourism Review, 66(3), 4e13.

97. Steiger, R., & Abegg, B. (2011). Climate change impacts on Austrian ski areas. In Borsdorf, J. Stötter, & E. Veulliet (Eds.), Proceedings of the Innsbruck conference November 21-23, 2011Managing Alpine future II (pp. 288e297). Vienna: Verlag der Österreichischen Akademie der Wissenschaften.

98. Steiger, R., & Stötter, J. (2013). Climate change impact assessment of ski tourism in Tyrol. Tourism Geographies, 1e24.

99. Steiger, R., & Trawöger, L. (2011). Klimawandel und Wintertourismus e Ein Vulnerabilitätsprofil für die Region Tirol. tw Zeitschrift für Tourismuswissenschaft, 3(2), 151e164.

100. Strobl, A., Steiger, R., Peters, M., & Weiermair, K. (2011). Tourismusforschung und epraxis: Wen kümmert der Klimawandel? In T. Bieger, C. Laesser, & P. Beritelli (Eds.), Schweizer Jahrbuch für Tourismus 2010Wettbewerb im alpinen Tourismus e Herausforderungen und Innovationen (pp. 105e114) Berlin: Erich Schmidt Verlag.

101. Sunstein, C. R. (2006). The availability heuristic, intuitive cost-benefit analysis, and climate change. Climatic Change, 77(1e2), 195e210.

102. Tervo, K. (2008). The operational and regional vulnerability of winter tourism to climate variability and change: the case of the

Finnish nature-based tourism entrepreneurs. Scandinavian Journal of Hospitality and Tourism, 8(4), 317e332.

103. Tervo-Kankare, K. (2011). The consideration of climate change at the tourism destination level in Finland: coordinated collaboration or talk about weather? Tourism Planning & Development, 8(4), 399e414.

104. Trawöger, L. (2010, August). Toward a vulnerability profile in winter tourism in Tyrol. In Paper presented at the 11th biennial conference of the international society for ecological economics (ISEE), Oldenburg.

105. Trawöger, L. (2011). Climate change and winter tourism e a stakeholder perspective in the Tyrol region. In A. Borsdorf, J. Stötter, & E. Veulliet (Eds.), Proceedings of the Innsbruck conference November 21e23, 2011Managing Alpine future II (pp. 332e339). Vienna: Verlag der Österreichischen Akademie der Wissenschaften.

106. Turton, S., Dickson, T., Hadwen, W., Jorgensen, B., Pham, T., immons, D., et al. (2010). Developing an approach for tourism climate change assessment: evidence from four contrasting Australian case studies. Journal of Sustainable Tourism, 18(3), 429e447.

107. Unbehaun, W., Pröbstl, U., & Haider, W. (2008). Trends in winter sport tourism: challenges for the future. Tourism Review, 63(1), 36e47.

108. UNWTO (United Nations World Tourism Organization). (2013). Tourism highlights. Madrid: World Tourism Organization.

109. Uyarra,M. C., Côté, I.M., Gill, J.A., Tinch, R. R., Viner,D.,&Watkinson, A. R. (2005). Islandspecific preferences of tourists for environmental features: implications of climate change for tourism-dependent states. Environmental Conservation, 32(1), 11e19.

110. Valls, J.-F., & Sardá, R. (2009). Tourism expert perceptions for evaluating climate change impacts on the Euro-Mediterranean tourism industry. Tourism Review, 64(2), 41e51.

111. Vanat, L. (2013). 2013 International report on snow & mountain tourism: Overview of the key industry figures for ski resorts. Retrieved from http://www.vanat.ch

112. Wachinger, G., Renn, O., Begg, C., & Kuhlicke, C. (2012). The risk perception paradox e implications for governance and communication of natural hazards. Risk Analysis, 33, 1049e1065.

113. Wagner, G., & Zeckhauser, R. J. (2012). Climate policy: hard problem, soft thinking. Climatic Change, 110(3e4), 507e521.

114. Weber, E. U. (2006). Experience-based and description-based perceptions of long-term risk:why globalwarming does not scare us (yet). Climatic Change, 77(1e2),103e120.

115. Weinstein, N. D. (1980). Unrealistic optimism about future life events. Journal of Personality and Social Psychology, 39(5), 806e820.

116. WKO (Wirtschaftskammer Österreich). (2012). Factsheet Seilbahnen. Retrieved from http://portal.wko.at.

117. Wolfsegger, C., Gössling, S., & Scott, D. (2008). Climate change risk appraisal in the Austrian ski industry. Tourism Review International, 12(1), 13e23.

Chapter 2

Comparison of Climate Preferences for Domestic and International Beach Holidays: A Case Study of Canadian Travelers

Michelle Rutty [1,*] and Daniel Scott [2]

[1] *Department of Community Sustainability, Michigan State University, East Lansing, MI 48824, USA*

[2] *Department of Geography & Environmental Management, University of Waterloo, Waterloo, ON N2L 3G1, Canada*

ABSTRACT

Coastal tourism is the largest segment of global leisure tourism and it is firmly linked to the destination's natural resources—with climatic resources chief among them. Through observations and survey responses of beach users, studies have evaluated climatic resources for coastal tourism by quantifying optimal and unacceptable conditions. However, these studies have not taken into consideration that different forms of holidays (e.g., daytrips, short trips, main annual holiday, "once-in-a-lifetime" trip) may have varying degrees of resilience to climatic conditions. This is the first study to explore whether ideal and unacceptable climatic conditions vary between domestic and international tourists. Using an in situ survey, Canadian beach users traveling domestically (n = 359) and internationally (n = 120) were examined. Key findings include statistically significant differences ($p \leq 0.05$) between the two sample groups for every climate variable, with the international sample more resilient to a broader range of weather conditions, including a greater acceptance for warm temperatures, longer rainfall durations, higher wind speeds, and greater cloud cover. This study adds further insight into the complexities of evaluating climate for tourism, with implications for the demand response of tourists to climate change.

Keywords: climate; tourism; tourist; beach; coastal; international; domestic; weather; Canada; Caribbean

1. INTRODUCTION

The cultural background of tourists and the influence this has on tourist behaviors and travel decisions has been highly researched. Extant literature demonstrates cross-cultural differences in travel motivations (push and pull), destination choice, activity selection, trip expectations, and satisfaction levels [1,2,3,4,5,6]. Although less researched, significant behavioral differences have also been recorded between domestic and international tourists. Evidence indicates that domestic tourists have different travel motivations compared to international tourists [7,8], as well as varying destination image perceptions [9], information search behavior and source preferences when trip-planning [10], recreational and shopping choices [11,12,13,14,15], service expectations [16,17], and degrees of destination loyalty and satisfaction [18]. The greater distances traveled by international tourists, including often-increased holiday length, are thought to be key factors that influence this behavioral difference from domestic visitors [13,15,19,20,21,22].

The decision to travel and the factors that guide destination choice is a complex behavioral process that involves an assessment of internal and external variables [23,24]. Weather and climate are important external factors that influence this process from the initial motivation to travel [7,25], to the timing and destination selected [7,26,27,28,29,30], to the overall travel experience itself [31,32,33,34,35]. While there has been much progress in understanding the significance of climatic resources for tourism demand, our knowledge of how tourists evaluate climatic resources, particularly those conditions that trigger a behavioral response, remains incomplete [36,37,38,39]. Specifically, existing research has not taken into consideration that different forms of holidays, such as daytrips, short trips, the main annual holiday, or the "once-in-a-lifetime" trip, may have varying degrees of resilience to climatic conditions [40,41]. An international trip with a high degree of planning and financial commitment prior to travel may result in different climatic preferences and tolerances compared to a spontaneous daytrip or short domestic trip.

It is important to understand how tourists evaluate climatic conditions and to quantify those conditions that are most preferred or avoided. Tourists have the greatest capacity to adapt to weather and climate, with the ability to alter their destination (i.e., change location), the timing of their trip (i.e., go another day), and the intensity (i.e., go less often) of their holidays, or substitute their intended travel activity for another (e.g., go on an urban sightseeing holiday instead of a winter sports holiday). The ease with which tourists can alter their travel plans is a challenge and concern for weather-sensitive tourism businesses, as well as destinations heavily dependent on tourism, as they cannot adapt as readily to variable climatic conditions due to large capital investments in immobile infrastructure (e.g., hotels, shopping facilities) [42]. The importance of this research is amplified with climate change, as a direct impact on tourism will be the global redistribution of climatic resources [37]. Climate change could change the length and quality of climate-sensitive tourism seasons, affecting the temporal and spatial distribution of international and domestic tourism flows and spending.

The aim of this study is to examine whether ideal and unacceptable climatic conditions for a beach trip vary between tourists traveling domestically and internationally. The analysis is based on primary data collected through in situ surveys of Canadian travelers visiting domestic Great Lakes beaches in the province of Ontario (Grand Bend, Kincardine, Toronto) (n = 359) and Canadians traveling internationally to visit beaches abroad in the Caribbean (Barbados, Saint Lucia, Tobago) (n = 120). The results from this study advance our understanding of tourists' climate needs, which is fundamental if we are to more accurately assess climate (change) impacts on tourism demand patterns (seasonally and geographically).

2. CLIMATE AS A RESOURCE FOR TOURISM

As indicated, the literature is clear that weather and climate represent a key element of the natural resource base of a tourism destination and this is an important consideration in tourist decision-making. Climate not only determines the suitability of a location for tourist activities, but it is the principal driver of seasonality in tourism demand—a defining characteristic of global tourism [43]. Given that leisure tourism is voluntary (i.e., people participate freely for personal satisfaction or pleasure), one must perceive the climate to be suitable for undertaking the types of activities desired at a specific destination, with an unfavorable climate inhibiting tourists' destination choice or range of activities pursued [44,45]. Therefore, the potential for tourists to travel to a destination is a function of the perceived appeal of the climate resources to tourists and their constraints on tourist activities, which can be classified along a spectrum from ideal to unacceptable [45,46,47].

Over the past three decades, a variety of methods have been used to quantify optimal and unacceptable climate conditions for a variety of types of tourism. Approaches include expert assessments, such as Besancenot et al.'s "weather typing" classification [48] and Miezkowski's integrative tourism climate index (TCI) [49]. Statistical models, including multiple linear regressions, have examined the effect of weather on visitation numbers at tourist attractions [50,51,52,53,54,55,56,57,58,59] and econometric models have estimated tourism-climate demand functions based on international tourism arrivals data at both a regional [60,61,62] and global scale [63,64]. Researchers have also observed tourists' in situ behavioral response to weather conditions (e.g., via webcams) [65,66,67] and have surveyed tourists directly about their preferred climatic conditions [25,29,38,42,68,69,70,71,72,73,74].

As the tourism climatology literature has developed, increasing evidence reveals the complexity of quantifying tourists' climatic preferences. Studies have shown that preferences vary based on the type of tourism environment or activity (e.g., beach, urban, mountain) [38,42,70,71,74], as well as interpersonal factors, such as socio-demographics (e.g., age, family status, gender) [38,70,75] and climatic region of origin [29,42,73]. The importance of psychological factors, including perceived control and expectations, can also significantly influence how tourists evaluate outdoor environments. For example, Rutty and Scott found beach tourists were willingly exposing themselves to (and preferred) very strong thermal stress (e.g., >38 °C) [38]. The thermal tolerance was linked to tourists' high level of autonomy, as they can choose the timing and location of their holiday, as well as the discretionary activities they participate in and the duration of exposure to the thermal conditions. This allows tourists to have a high level of thermal control, whereby they can either adjust their thermal conditions (e.g., move into the shade or go swimming) or leave the beach when the conditions are no longer personally suitable. Climatic preferences have also been shown to vary based on seasonal expectations, with people preferring higher temperatures at their place of residence in the summer compared to the winter season [76,77,78]. Natural seasonality is a primary driver of some of the world's largest global tourism flows (e.g., northern Europe to southern Europe/Mediterranean, northern USA and Canada to southern USA/Caribbean), with tourists investing time and money to travel long distances to escape cooler temperate regions and experience conditions that are significantly warmer than at home. It is therefore possible that tourists' climatic preferences vary depending on whether or not the beach holiday is domestic or international.

3. METHODS

This study explores the climatic preferences of Canadians, specifically Canadian coastal tourists traveling internationally to the Caribbean as well as domestically within the province of Ontario (Canada). Coastal tourism is the largest segment of global leisure tourism [79,80,81,82] and it is firmly linked to the destination's natural resources—with climatic resources chief among them [7,29,46,83,84]. Multiple studies provide evidence of the sensitivity of coastal tourists to weather conditions. For example, low temperatures, high winds and precipitation have been linked to low levels of beach use, with high temperatures and sunshine linked to strong levels of demand [65,66,67,85]. Major intra-regional tourism demand patterns highlight the influence of seasonal climate for coastal tourism. Millions of North Americans travel south to the warm and sunny coasts of the Caribbean during the winter months (December to April), including 2.5 million Canadians, annually [86]. In Ontario, the province received over 127 million domestic tourists in 2012, with the largest share (32%) arriving during the summer months (July–September) [87].

The results from this study are based on primary data collected through in situ surveys of Canadian beach users. The self-administered surveys were disseminated during peak beach usage time (11 a.m. to 5 p.m. local time), with all beach users who were sitting or lying down on the sand approached to participate. The international sample (i.e., Canadians traveling abroad) was collected over three weeks between March and April 2012 at Caribbean beaches located alongside resorts on the islands of Barbados (Accra, Amaryllis, Dover, Holetown), Saint Lucia (Gros Islet, Rodney Bay) and Tobago (Crown Point, Pigeon Point) (n = 120). The domestic sample (i.e., Canadians traveling in Ontario) was collected over three weeks between July and August 2012 at beach-oriented parks on the Great Lakes, including Grand Bend, Kincardine, and Toronto (n = 359). The survey was conducted in English in all of the locations, with a high response rate of 85% or greater at each beach.

Respondents were asked about their preferred and unacceptable climatic conditions for their beach holiday, which included temperature, rain, wind and cloud cover. For temperature preferences and thresholds, respondents were asked to circle the range of temperature(s) they deemed ideal and unacceptable for their beach holiday. For the variables of rain, wind and cloud cover, respondents were asked to select the most preferred and unacceptable daily conditions from a list of five available options (i.e., multiple choice). The survey used in this study builds on previous climate surveys of tourists [29,42,71,73], facilitating needed comparisons across studies.

Given the geographic size of Canada, the latitudinal origin may impact tourist's climatic preferences. To control for this possible bias, the sample does not include Canadians residing in continental subarctic, boreal or tundra climate regions of Canada. There is, however, the possibility of sampling bias with an in situ survey, as noted in previous studies [42,73]. First, beach users with preferences for weather conditions that are not available during the sampling period may not be represented (i.e., if the weather conditions were deemed unsuitable to be on the beach, that respondent would not be on site to be surveyed). With this consideration in mind, the sampling period occurred during the tourism high season in both the domestic and international locations to capture the demand preferences of the majority. It is also important to note that the weather conditions during both sampling periods were representative of the long-term averages for the two regions. A second potential limitation relates to the distribution of surveys to beach users on the "dry" part of the beach. Tourists in the water at the time of the survey were not approached to participate, which may result in an under-representation of those who more often engage in water activities. The extent of sampling bias due to these two limitations is unknown.

4. RESULTS

4.1. Respondent Characteristics

In both samples, there were more females than males, with 20% more females in the international sample and 34% more females in the domestic sample. The international sample was older, particularly with respect to tourists aged 55 years and above (39% of the total sample vs. 12%). Conversely, the domestic sample had a larger percentage of respondents that were between the ages of 18–24 (20% vs. 8%) and 25–34 (22% vs. 13%). This age difference reflects the demographics of tourists at the study area beaches, with an older and wealthier demographic traveling to Caribbean destinations in winter months. The international sample had more travel experience at the survey location, with almost all of the respondents having traveled to the Caribbean before (92%), including more than half (59%) who had traveled to the Caribbean six or more times. For the domestic sample, the majority had traveled to the Ontario study location at least once before (83%), but less than half had traveled to the location six or more times (44%).

4.2. Climate Preferences

When respondents were asked about their preferred temperatures for a beach holiday, the largest share (30%) of the international sample indicated that they prefer temperatures between 27 and 30 °C for their Caribbean beach holiday, which is slightly warmer than the 25–30 °C that is preferred for a domestic beach holiday ($\chi(1) = 36.324$, $p = 0.01$) (Figure 1). Statistically significant differences were also recorded for unacceptably cool thresholds ($\chi(1) = 69.803$, $p = 0.000$). The majority (50%) of the Canadians traveling in the Caribbean stated that <23 °C would be too cool for a beach holiday compared to <21 °C for Canadians traveling domestically. Unacceptably hot temperatures differed by 1 °C between the two samples, with Canadians traveling in the Caribbean more accepting of high temperatures compared to Canadians travelling in Ontario (34 °C and 33 °C, respectively), but this difference was not statistically significant ($p = 0.087$).

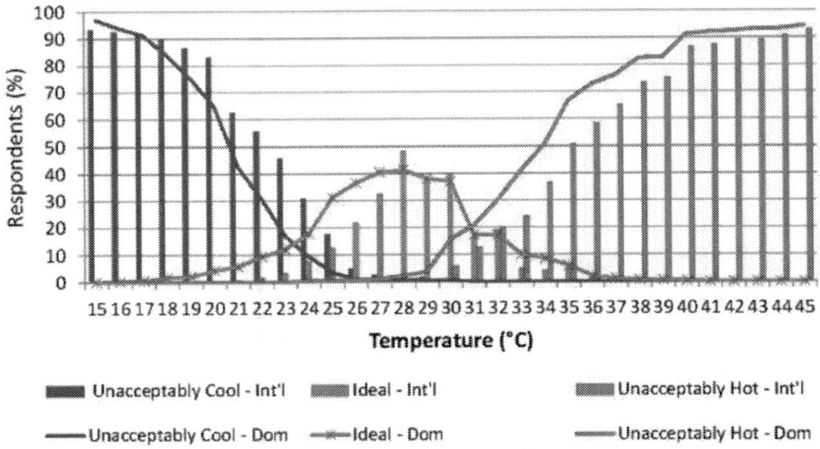

Figure 1. Unacceptably cool (blue), ideal (green), and unacceptably hot (red) temperatures for Canadian beach tourists traveling internationally (stacked bar) and domestically (line).

Statistically significant differences were also recorded for rain preferences (χ(1) = 211.839, p = 0.000). The majority of domestic beach tourists indicated no rain as ideal for their beach holiday (91%), while the largest share of Canadians traveling abroad indicated ≤15 min of rain as ideal (48%) (Table 1). Unacceptable rain conditions were similarly significantly different, with domestic travelers less accepting of all rain conditions greater than 15 min compared to Canadian travelers abroad (≤15 min, χ(1) = 57.777, p = 0.000; 30 min–1 h, χ(1) = 83.367, p = 0.000; 2–4 h, χ(1) = 26.443, p = 0.000).

Table 1. Ideal and unacceptable daily rain conditions (% of respondents).

	No Rain	≤15 min	30 min–1 h	2–4 h	≥5 h
Ideal					
Domestic	91	5	2	1	1
International	28	48	24	0	0
Unacceptable					
Domestic	3	43	43	83	94
International	10	5	27	60	97

Statistically significant differences were also recorded for wind preferences ($\chi(1) = 30.647$, $p = 0.000$), with domestic tourists preferring less wind while at the beach (93% stated no wind or a light breeze) and Canadians traveling internationally preferring windier conditions (99% stated a light breeze or moderate wind) (Table 2). Canadians traveling abroad were statistically less accepting of no winds ($\chi(1) = 4.864$, $p = 0.027$) and more accepting of higher wind speeds, particularly with respect to moderate (30% vs. 45%) ($\chi(1) = 8.1C5$, $p = 0.004$) and strong (83% vs. 75%) wind speeds, although the latter was not statically significant ($p = 0.060$).

Table 2. Ideal and unacceptable daily wind conditions (% of respondents).

	No Wind	Light Breeze	Moderate Wind	Strong Wind	Very Strong Wind
Ideal					
Domestic	9	84	7	0	0
International	1	77	22	0	0
Unacceptable					
Domestic	11	2	45	83	94
International	19	1	30	75	93

Although the majority of both sample groups preferred 25% cloud cover for their domestic or international beach holiday (54% and 76%, respectively), domestic tourists preferred less cloud overall when compared to the international sample group ($\chi(1) = 24.393$, $p = 0.000$) (Table 3). For unacceptable cloud conditions, the majority of both domestic and international beach tourists indicated 75% cloud cover as the threshold of acceptability (64% and 65%, respectively), with no statistically significant differences recorded ($p \geq 0.000$).

Table 3. Ideal and unacceptable cloud cover (% of respondents).

	Percent of Cloud Cover				
	0	25	50	75	100
Ideal					
Domestic	35	54	9	2	0
International	13	76	11	0	0
Unacceptable					
Domestic	4	5	32	65	89
International	7	2	23	64	89

5. DISCUSSION

The results from this research indicate that tourists' climatic preferences and thresholds for a beach holiday can differ depending on whether the holiday is domestic or international. Canadians traveling in Ontario for a beach holiday prefer and are more accepting of cooler conditions compared to when they travel abroad, which is when they prefer and are more accepting of warmer conditions. The ideal temperature conditions identified by the sample abroad (27–30 °C) are the same or within one degree of the recorded beach preferences identified in other tourism surveys, while the ideal temperatures for the domestic sample (25–30 °C) are up to 4 °C cooler [29,42,71,73]. The cooler temperatures preferred by the domestic sample in this study can likely be explained by the fact that previous studies have all examined international tourists that are effectively paying to escape cooler conditions to enjoy a warm (hot) holiday (i.e., the survey sample has included tourists traveling from Canada to the Caribbean, or Northern and Central Europe to the Mediterranean), whereas the domestic sample in this study is traveling during the warm summer season to experience similar climatic conditions to their place of origin or perhaps to escape the urban heat by travelling to cooler lake-side conditions. Thermal comfort theory suggests that people adjust their climatic preferences based on expectations, with evidence that the same individual will have differing temperature preferences depending on their expectations of the outdoor conditions [77,78,88]. Tourists leaving temperate regions for tropical holidays are expecting to experience warm/hot conditions. High temperature expectations are not only well-formed given that the vast majority of tourists gather weather information prior to their trip [26,71,89,90,91], but also because of the prominence of weather and climate in marketing the image of coastal destinations such as the Caribbean [46,92,93].

In terms of rain, Canadians traveling abroad are more accepting of rain compared to Canadians traveling domestically. As noted, the international sample has more travel experience with a Caribbean beach holiday than the domestic sample has with an Ontario beach holiday. Perhaps the international tourists understand that the Caribbean dry season weather (i.e., January–June) is dominated by convective storms, with rainfall typically falling in small amounts (e.g., average precipitation in Barbados during the study period is 50 mm; St. Lucia 45 mm; Tobago 40 mm) and for a short duration of time before the skies clear. Similar rain acceptances have also been recorded for tourists from tropical regions who would be similarly experienced with such weather patterns in the region [73]. In Ontario, summer rainfall amounts (i.e., June–August) are not only greater (e.g., average precipitation in Grand Bend during the study is 89 mm; Kincardine 77 mm; Toronto 70 mm), but it is also more likely to persist for a longer duration of time, including longer periods with cloud cover before and after the rain event.

Although both the domestic and international beach tourists had the same ideal and unacceptable wind preferences (light wind and strong winds, respectively), the former was statistically more accepting of no wind and the latter more accepting of higher winds. The influence of wind speed on cooling the body may be linked to the preferences recorded in this study. The higher winds preferred by the Canadians traveling to the Caribbean would reduce the heat load experienced at the higher ideal temperatures (27–30 °C vs. 25–30 °C), whereas no wind would enhance thermal comfort at the lower temperatures preferred by the domestic tourists, particularly with respect to acceptably cool conditions (>21 °C vs. >23 °C). Similarly for cloud cover, domestic tourists preferred less cloud, which would increase the intensity of thermal radiation and thereby increase thermal comfort at cooler temperatures. These findings suggest that tourists can distinguish the influencing effect of multiple weather parameters, underscoring the need to assess multiple climatic resources for tourism because tourists experience and respond to these combined influences in thermo-physiologically relevant ways [38,45].

6. CONCLUSIONS

Jafari [19] hypothesized that once an individual arrives at a destination, home life assumes a residual position as the tourist sheds the culture of their home environment and assumes a tourist culture. In so doing, the tourist becomes less critical and more tolerant of the new environment. Since domestic tourists are in an environment that is similar to that of their place of origin, these similarities may prevent them from adopting this less critical behavior. Studies have found that distance traveled, spatial change, and length of holiday are mportant factors that influence behavioral differences between international and domestic tourists [13,15,20,21,22]. Despite both sample groups comprising tourists of the same nationality and climatic region of origin (i.e.,

Canada), ideal and unacceptable climatic conditions for a beach holiday differed for all variables examined.

This study empirically supports the assumptions of Hall [40] and Scott and Lemieux [41] that different forms of holidays, such as a short domestic trip compared to a long-haul international holiday, can have varying degrees of resilience to climatic conditions. Canadians traveling abroad were more resilient to a broader range of weather conditions, including a greater acceptance for warm temperatures, longer rainfall durations, higher wind speeds, and cloud cover. Unlike an international trip, which often requires a high degree of financial commitment and planning prior to departure, a domestic tourist has the ability to plan their trip with minimal or no prior commitment. Domestic tourists are therefore more flexible and can afford to accept a narrower range of climatic conditions and decide last minute whether or not to travel to the beach based on the weather forecast. This last-minute decision-making is supported by how the two sample groups gathered weather information when planning their trip; 57% of the Canadians traveling abroad gathered weather information before booking their flights or accommodation, whereas only 36% of the domestic sample gathered weather information three or more days in advance of their trip. Also, on the day of the survey, 57% of the domestic sample looked at the weather information, whereas less than half (40%) of the Canadians abroad looked at the forecast.

This study also adds further insight into the complexities of evaluating climate for tourism. Growing research reveals that ideal and unacceptable climatic conditions differ depending on the specific tourism environments or activities, as well as on personal factors, including age, family status, gender, and nationality [29,38,42,70,73,74,75]. The results from this study are the first to reveal that it is also important to consider whether or not the holiday is domestic or international, as this can also influence tourist behavior and travel decisions. Studies that have broadly generalized tourism climate resources (optimal and unacceptable conditions) (e.g., weather typing, climate indices, multiple linear regression models, econometric demand models) do not adequately capture the complexity of tourists' climatic preferences or the implications of diverse preferences for tourist decision-making.

Collectively, this research has implications for the demand response of tourists to climate change. A direct impact of climate change on tourism will be the global redistribution of climatic resources, which will affect the temporal and spatial distribution of domestic and international tourism flows and economic spending [94]. An increasing number of publications have sought to analyze how travel flows could be altered by climate change [57,60,61,63,64,95,96,97,98,99,100,101]. The results from these publications are meant to identify potentially vulnerable destinations and assist the tourism industry in planning future operations, including the development of adaptation strategies to minimize associated risks and capitalize on new opportunities posed by changes in the competitive relationships among destinations [98]. However, such assessments have not been without criticism, for they do not account for the complex role of weather and climate in tourist behavior and decision-making [36,37,46,102,103]. Understanding how tourists evaluate climatic resources, including the contextual influences that guide this response, is a prerequisite if accurate projections are to be made about changes in tourism demand as a result of climate variability or future climate change. With only a handful of studies examining how climatic parameters are evaluated and perceived by tourists, more research is needed to ascertain detailed tourist weather sensitivities across broader tourism environments and activities. This includes both studies that examine the interaction between trip type (e.g., domestic, international, short-haul, long-haul) and climatic preferences, as well as the influence of other socio-demographic factors (e.g., age, family status) on climatic preferences. In so doing, it will allow needed comparisons among different tourism market segments and provide further insight into the influence of climate change on tourism demand patterns.

ACKNOWLEDGMENTS

Financial support from the Canada Research Chairs program, Social Sciences and Humanities Research Council of Canada (SSHRC), and the International Development Research Centre (IDRC) are gratefully acknowledged.

AUTHOR CONTRIBUTIONS

Both authors conceived and designed the experiments, as well as analyzed the data; Michelle Rutty performed the experiments and wrote the paper.

CONFLICTS OF INTEREST

The authors declare no conflict of interest.

ABBREVIATIONS

The following abbreviation is used in this manuscript:

TCI Tourism Climate Index

REFERENCES

1. Pizam, A.; Sussman, S. Does nationality affect tourist behavior? Ann. Tour. Res. 1995, 22, 901–917.
2. Crotts, J.; Erdmann, R. Does national culture influence consumers' evaluation of travel services? A test of Hofstede's model of cross-cultural differences. Manag. Serv. Qual. 2000, 10, 410–419.
3. Reisinger, Y.; Turner, L.W. Cross-Cultural Behavior in Tourism: Concepts and Analysis; Butterworth Heinemann: Oxford, UK, 2003.
4. Lee, C.-K.; Lee, Y.-K.; Wicks, B.E. Segmentation of festival motivation by nationality and satisfaction. Tour. Manag. 2004, 25, 61–70.
5. Prayag, G.; Ryan, C. The relationship between the "push" and "pull" factors of a tourist destination: The role of nationality—An analytical qualitative research approach. Curr. Issues Tour. 2011, 14, 121–143.
6. Correia, A.; Kozak, M.; Ferradeira, J. Impact of culture on tourist decision-making styles. Int. J. Tour. Res. 2011, 13, 433–446.
7. Kozak, M. Comparative analysis of tourist motivation by nationality and destinations. Tour. Manag. 2002, 23, 221–232.
8. Kim, K. Understanding differences in tourist motivation between domestic and international travel: The university student market. Tour. Anal. 2007, 12, 65–75.

9. Bonn, M.A.; Joseph, S.M.; Dai, M. International versus domestic visitors: An examination of destination image perceptions. J. Travel Res. 2005, 43, 294–301.

10. Ortega, E.; Rodriguez, B. Information at tourism destinations. Importance and cross-cultural differences between international and domestic tourists. J. Bus. Res. 2007, 60, 146–152.

11. Wang, Z.H.; Ryan, C. New Zealand retailers' perceptions of some tourists'negotiation styles for souvenir purchases. Tour. Cult. Commun. 1998, 1, 139–152.

12. Field, A.M. The college student market segment: A comparative study of travel behaviors of international and domestic students at a southern university. J. Travel Res. 1999, 37, 375–381.

13. Caar, N. A comparative analysis of the behavior of domestic and international young tourists. Tour. Manag. 2002, 23, 321–325.

14. Kim, K.; Jogaratnam, G. Activity preferences of Asian international and domestic American university students: An alternate basis for segmentation. Vacat. Mark. 2003, 9, 260–270.

15. Yuksel, A. Shopping experience evaluation: A case of domestic and international visitors. Tour. Manag. 2004, 25, 751–759.

16. Mok, C.; Armstrong, W.R. Expectations for hotel service quality: Do they differ from culture to culture? J. Vacat. Mark. 1998, 4, 381–391.

17. Lloyd, A.E.; Yip, L.S.C.; Luk, S.T.K. An examination of the differences in retail service evaluation between domestic and tourist shoppers in Hong Kong. Tour. Manag. 2011, 32, 520–533.

18. Mechinda, P.; Serirat, S.; Gulid, N. An examination of tourists' attitudinal and behavioral loyalty: Comparison between domestic and international tourists. J. Vacat. Mark. 2009, 15, 129–148.

19. Jafri, J. Tourism models: The sociocultural aspects. Tour. Manag. 1987, 8, 151–159.

20. Hoxter, A.L.; Lester, D. Tourist behavior and personality. Personal. Individ. Differ. 1998, 9, 177–178.

21. Laing, A.N. The Package Holiday: Participant, Choice and Behavior. Unpublished Ph.D. Thesis, Hull University, Yorkshire, UK, 1987.

22. Debbage, K. Spatial behavior in a Bahamian resort. Ann. Tour. Res. 1991, 18, 251–268.

23. Woodside, A.G.; Lysonski, S. A general model of traveler destination choice. J. Travel Res. 1989, 16, 14–18.

24. Um, S.; Crompton, J.L. Attitude determinants in tourism destination choice. Ann. Tour. Res. 1990, 17, 432–448.

25. Lohmann, M.; Kaim, E. Weather and holiday preference—Image, attitude and experience. Rev. Tour. 1999, 2, 54–64.

26. Hamilton, J.M.; Lau, M.A. The role of climate information in tourist destination choice decision-making. In Tourism and Global Environmental Change; Gössling, S., Hall, C.M., Eds.; Routledge: London, UK, 2005.

27. Eugenio-Martin, J.L.; Campos-Soria, J.A. Climate in the region of origin and destination choice in outbound tourism demand. Tour. Manag. 2010, 31, 744–753.

28. Hill, A. Holiday deal abroad vanish in rush to flee the rain. The Observer, 9 August 2009.

29. Moreno, A. Mediterranean tourism and climate (change): A survey-based study. Tour. Hosp. Plan. Dev. 2010, 7, 253–265.

30. Hadwen, W.; Arthington, A.; Boon, P.; Taylor, B.; Fellows, C. Do climatic or institutional factors drive seasonal patterns of tourism visitation to Protected Areas across diverse climate zones in Eastern Australia? Tour. Geogr. 2011, 13, 187–208.

31. Bardón, F.E. Resumen del Estudio Sobre el Grado de Satisfaccio'n de la Demanda Turistica Nacional y Extranjera en Relacion con el Product Turistico Espanol. Ministerio de

Transportes; Turismo y Comunicaciones: Madird, Spain, 1991.

32. Williams, P.; Dossa, K.; Hunt, J. The influence of weather context on winter resort evaluations by visitors. J. Travel Res. 1997, 36, 29–36.

33. Becken, S.; Wilson, J.; Reisinger, A. Weather, Climate and Tourism: A New Zealand Perspective; Land environment and people—Technical Report 20; Lincoln University: Christchurch, New Zealand, 2010.

34. Becken, S.; Wilson, J. The impacts of weather on tourist travel. Tour. Geogr. 2013, 15, 620–639.

35. Tervo-Kankare, K.; Hall, M.C.; Saarinen, J. Christmas tourists' perceptions to climate change in Rovaniemi, Finland. Tour. Geogr. 2013, 15, 292–317.

36. Gössling, S.; Scott, D.; Hall, M.C.; Ceron, J.-P.; Dubois, G. Consumer behavior and demand response of tourists to climate change. Ann. Tour. Res. 2012, 39, 36–58.

37. Scott, D.; Gössling, S.; Hall, C.M. International tourism and climate change. WIREs Clim. Chang. 2012, 3, 213–232.

38. Rutty, M.; Scott, D. Bioclimatic comfort and the thermal perceptions and preferences of beach tourists. Int. J. Biometeorol. 2015, 59, 37–45.

39. Rutty, M.; Scott, D.; Johnson, P.; Jover, E.; Pons, M.; Steiger, R. Behavioral adaptation of skiers to climatic variability and change in Ontario, Canada. J. Outdoor Recreat. Tour. 2015, 11, 13–21.

40. Hall, C.M. Tourism: Rethinking the Social Science of Mobility; Prentice-Hall: Harlow, UK, 2005.

41. Scott, D.; Lemieux, C. Weather and climate information for tourism. Procedia Environ. Sci. 2010, 1, 146–183.

42. Scott, D.; Gössling, S.; de Freitas, C. Preferred climates for tourism: Case studies from Canada, New Zealand and Sweden. Clim. Res. 2008, 38, 61–73.

43. Butler, R. Seasonality in tourism: Issues and implications. In Seasonality in Tourism; Baum, T., Lundtorp, S., Eds.; Elsevier: London, UK, 2001.

44. Smith, K. The influence of weather and climate on recreation and tourism. Weather 1993, 48, 398–404.

45. De Freitas, C.R. Recreation climate assessment. Int. J. Climatol. 1990, 10, 89–103.

46. Gomez-Martin, M.A. Weather, climate and tourism a geographical perspective. Ann. Tour. Res. 2005, 32, 571–591.

47. Scott, D.; Hall, C.M.; Gössling, S. Tourism and Climate Change: Impacts, Adaptation and Mitigation; Routledge: London, UK, 2012.

48. Besancenot, J.P.; Mouiner, J.; de Lavenne, F. Les conditions climatiques du tourisme, littoral. Norois 1978, 99, 357–382.

49. Mieczkowski, Z. The tourism climate index: A method of evaluating world climates for tourism. Can. Geogr. 1985, 29, 220–233.

50. Van Lier, H.N. Determination of Planning Capacity and Layout Criteria of Outdoor Recreation Projects; Agriculture research reports 705; Centre for Agricultural Publishing and Documentation: Wageningen, The Netherlands, 1973.

51. Gibbs, K.C.; McGuire, J.F. Estimation of Outdoor Recreation Values; Economic Report 53; Food and Resource Economics Department, Agricultural Experiment Station Institute of Food and Agricultural Science: Gainesville, FL, USA, 1973.

52. Emmons, H.; Dean, B.V.; Nunnikhoven, T.S.; Rossiter, D.R.; Rao, R.N.; de Kluyver, C.; Richard, B. A market analysis for Cleveland Zoological Park. In Management Science Applications to Leisure Time Operations; Lodamy, S.P., Ed.; Elsvier: New York, NY, USA, 1975.

53. Tolley, G.; Randall, A.; Blomquist, G.; Brien, M.; Fabian, R.; Fishelson, G.; Frankel, A.; Grenchik, M.; Hoehn, J.; Kelly, A.; et al. Establishing and Valuing the Effects of Improved Visibility in Eastern United States; United States Environmental Protection Agency: Chicago, IL, USA, 1986.

54. Meyer, D.; Dewar, K. A new tool for investigating the effect of weather on visitor numbers. Tour. Anal. 1999, 4, 145–155.

55. Jones, B.; Scott, D. Climate change, seasonality and visitation to Canada's National Parks. J. Parks Recreat. Adm. 2006, 24, 42–62.

56. Scott, D.; Jones, B. A regional comparison of the implications of climate change on the golf industry in Canada. Can. Geogr. 2007, 51, 219–232.

57. Nicholls, S.; Amelung, B. Climate change and tourism in northwestern Europe: Impacts and adaptation. Tour. Anal. 2008, 13, 21–31.

58. Serquet, G.; Rebetez, M. Relationship between tourism demand in the Swiss Alps and hot summer air temperatures associated with climate change. Clim. Chang. 2011, 108, 291–300.

59. Day, J.; Chin, N.; Sydnor, S.; Cherkauer, K. Weather, climate, and tourism performance: A quantitative analysis. Tour. Manag. Perspect. 2013, 5, 51–56.

60. Maddison, D. In search of warmer climates? The impact of climate change on flows of British tourists. Clim. Chang. 2001, 49, 193–208.

61. Lise, W.; Tol, R.S.J. Impact of climate on tourist demand. Clim. Chang. 2002, 55, 429–449.

62. Hamilton, J.M. Climate and the Destination Choice of German Tourists; Working Paper Number 2 (FNU-15); Research Unit Sustainability and Global Change: Hamburg, Germany, 2003.

63. Bigano, A.; Hamilton, J.M.; Tol, R.S.J. The impact of climate on holiday destination choice. Clim. Chang. 2006, 76, 389–406.

64. Berrittella, M.; Biagno, A.; Roson, R.; Tol, R.S.J. A general equilibrium analysis of climate change impacts on tourism. Tour. Manag. 2006, 27, 913–924.

65. Moreno, A.; Amelung, B.; Santamarta, L. Linking beach recreation to weather conditions: A case study in Zandvoort, Netherlands. Tour. Mar. Environ. 2009, 5, 111–119.

66. Martinez-Ibarra, E. The use of webcam images to determine tourist–climate aptitude: Favourable weather types for sun and beach tourism on the Alicante coast (Spain). Int. J. Biometeorol. 2011, 55, 373–385.

67. Gomez-Martin, M.A.; Martinez-Ibarra, E. Tourism demand and atmospheric parameters: Non-intrusive observation techniques. Clim. Res. 2012, 51, 135–145.

68. Morgan, R.; Gatell, E.; Junyent, R.; Micallef, A.; Ozhan, E.; Williams, A. An improved user-based beach climate index. J. Coast. Conserv. 2000, 6, 41–50.

69. Gomez-Martin, M.A. Climate potential and tourist demand in Catalonia (Spain) during the summer season. Clim. Res. 2006, 32, 75–87.

70. Credoc. Climat, Meteorology et Frequentation Touristique, Rapport Final; Ministere de L'Ecologie, de l'Eergie, du Developpement durable et de la Mer: Paris, France, 2009.

71. Rutty, M.; Scott, D. Will the Mediterranean become "too hot" for tourism? A Reassessment. Tour. Hosp. Plan. Dev. 2010, 7, 267–281.

72. Denstadli, J.M.; Jacobsen, J.K.; Lohmann, M. Tourists perceptions of summer weather in Scandinavia. Ann. Tour. Res. 2011, 38, 920–940.

73. Rutty, M.; Scott, D. Differential climate preferences of international beach tourists. Clim. Res. 2013, 57, 259–269.

74. Hewer, M.J.; Scott, D.; Gough, W.A. Tourism climatology for camping: A case study of two Ontario parks (Canada). Theory Appl. Climatol. 2015, 121, 401–411.

75. Limb, M.; Spellman, G. Evaluating domestic tourists' attitudes to British weather. A qualitative approach. In Proceedings of the 1st International Workshop on Climate, Tourism and Recreation, Halkidiki, Greece, 5–10 October 2001.

76. Nikolopoulou, M.; Baker, N.; Steemers, K. Thermal comfort in outdoor urban spaces: Understanding the human parameter. Sol. Energy 2001, 70, 227–235.

77. Spagnolo, J.; de Dear, R.J. A field study of thermal comfort in outdoor and semi-outdoor environments in subtropical Sydney Australia. Build. Environ. 2003, 38, 721–738.

78. Lin, T.P. Thermal perception, adaptation and attendance in a public square in hot and humid regions. Build. Environ. 2009, 44, 2017–2026.

79. Hall, C.M. Trends in ocean and coastal tourism: The end of the last frontier? Ocean Coast. Manag. 2001, 44, 601–618.

80. Honey, M.; Krantz, D. Global Trends in Coastal Tourism. Centre on Ecotourism and Sustainable Development, 2007. Available online: http://responsibletravel.org/resources/documents/reports/Global_Trends_in_Coastal_Tourism_by_CESD_Jan_08_LR.pdf (accessed on 3 February 2016).

81. United Nations Environmental Program (UNEP). Sustainable Coastal Tourism. An Integrated Panning and Management Approach, 2009. Available online: http://www.unep.fr/shared/publications/pdf/DTIx1091xPA-SustainableCoastalTourism-Planning.pdf (accessed on 3 February 2016).

82. Jones, A.; Phillips, M. Disapperaing Destinations. Climate Change and Future Challenges for Coastal Tourism; The Centre for Agriculture and Bioscience International (CABI): Wallingford, UK, 2011.

83. Mansfeld, Y.; Freundlich, A.; Kutiel, H. The relationship between weather conditions and tourists' perceptions of comfort: The case of the winter sun resort of Eilat. In Climate Change and Tourism—Assessment and Coping Strategies; Amelung, B., Blazejczyk, K., Matzarakis, A., Eds.; Institute of Geography and Spatial Organization PAS: Maastricht, The Netherlands, 2004.

84. Uyarra, M.C.; Cote, I.M.; Gill, J.A.; Tinch, R.R.T.; Viner, D.; Watkinson, A.R. Island-specific preferences of tourists for environmental features: Implications of climate change for

tourism-dependent states. Environ. Conserv. 2005, 32, 11–19.

85. De Freitas, C.R. Tourism climatology: Evaluating environmental information for decision making and business planning in the recreation and tourism sector. Int. J. Biometeorol. 2003, 48, 45–54.

86. Statistics Canada. International Travel, 2013. Available online: http://.statcan.gc.ca/pub/66–201-x/2010000/part-partie1-eng.htm (accessed on 15 October 2015).

87. Ontario Ministry of Tourism, Culture & Sport. Ontario Regional Tourism Profile, 2012. Available online: http://www.mtc.gov.on.ca/en/research/rtp/rtp.shtml (accessed on 15 October 2015).

88. Andrade, H.; Alcoforado, M.-J.; Oliveira, S. Perceptions of temperature and wind by users of public outdoor spaces: Relationships with weather parameters and personal characteristics. Int. J. Biometeorol. 2011, 55, 665–680.

89. Gamble, D.W.; Leonard, L.A. Coastal Climatology Products for Recreation and Tourism End Users in Southeastern North Carolina; NOAA Coastal Services Centre: Charleston, SC, USA, 2005.

90. Hübner, A.; Gössling, S. Tourists perceptions of extreme weather events in Martinique. J. Destin. Mark. Manag. 2012, 1, 47–55.

91. Rutty, M.; Andrey, J. Weather forecast use for winter recreation. Weather Clim. Soc. 2014, 6, 293–306.

92. Besancenot, J. Clima y Turismo; Masson: Barcelona, Spain, 1991.

93. Perry, A. Climate and weather information for the package holiday-maker. Weather 1993, 48, 410–414.

94. Scott, D.; Amelung, B.; Becken, S.; Ceron, J.P.; Dubois, G.; Gössling, S.; Peeters, P.; Simpson, M. Climate Change and Tourism: Responding to Global Challenges; World Tourism Organization: Madrid, Spain, 2008.

95. Scott, D.; McBoyle, G.; Schwartzentruber, M. Climate change and the distribution of climatic resources for tourism in North America. Clim. Res. 2004, 27, 105–117.

96. Amelung, B.; Viner, D. Mediterranean tourism: Exploring the future with the tourism climate index. J. Sustain. Tour. 2006, 14, 349–366.

97. Amelung, B.; Nicholls, S.; Viner, D. Implications of global climate change for tourism flows and seasonality. J. Travel Res. 2007, 45, 285–296.

98. Hein, L.; Metzger, M.J.; Moreno, A. Potential impacts of climate change on tourism; a case study for Spain. Curr. Opin. Environ. Sustain. 2009, 1, 170–178.

99. Moore, W.R. The impact of climate change on Caribbean tourism demand. Curr. Issues Tour. 2010, 13, 495–505.

100. Perch-Nielsen, S.; Amelung, B.; Knutt, R. Future climate resources for tourism in Europe based on the daily Tourism Climate Index. Clim. Chang. 2010, 103, 363–381.

101. Hamilton, J.; Maddison, D.; Tol, R.S.J. Climate change and international tourism: A simulation study. Glob. Environ. Chang. 2005, 15, 253–266.

102. Gössling, S.; Hall, C.M. Uncertainties in predicting tourist flows under scenarios of climate change. Clim. Chang. 2006, 79, 163–173.

103. Gössling, S.; Hall, C.M. Uncertainties in predicting travel flows: Common ground and research needs. A reply to Bigano et al. Clim. Chang. 2006, 79, 181–183.

Chapter 3

Wolin Island, Tourism and Conceptions of Identity ☆

Barbara Bożętka

Gdańsk University, Institute of Geography, 4 J. Bażyńskiego Street, 80-952 Gdańsk, Poland

ABSTRACT

The study focuses on conceptions of the identity of Wolin Island, the area located in the southern part of the Baltic Sea. Identity is regarded as a complex phenomenon, and the research incorporates two primary perspectives: socio-cultural and spatial. Investigations reveal that key factors affecting the identity of the island include tourism development and a post-war history connected with significant migration processes. In consequence, the relationships between tourism and identity have been emphasised.

Investigation of the theme is accompanied by an analysis of the role of external elements posing a threat to Wolin's identity. They include worldwide factors, which embrace a cultural shift, devaluation of landscape or place meaning, and more site-specific processes, such as the local growth of the heritage industry. The discussion is followed by reflections on intrinsic features of identity and its evolution. In conclusion, main elements for construction of Wolin Island's identity are distinguished, and the lack of dependence on 'islandness' is noted.

Keywords: Identity; Tourism; Island; Northern Poland; The Baltic Sea

INTRODUCTION

Narratives of identities usually reveal many layers of meaning. Islands constitute especially valuable fields of inquiry, they demonstrate tremendous diversity with landscapes ranging from arctic deserts or tropical forests to compact urbanised areas, and simultaneously exhibit great cultural differences. Connected with remoteness, isolation and to a large degree with uniqueness, islands form the basis of manifold captivating images. However, there can be a sharp contradiction between reality and external associations. It is intriguing that questions regarding islands' identities frequently focus attention on their perception by 'outsiders'. Though offering considerable advantages, such an approach cannot reveal internal notions of identity, which are inseparable from inherent features of an island, including its socio-cultural disposition.

The study presented here investigates the identity of Wolin, an island situated in the southern part of the Baltic Sea. The work concentrates on Wolin's distinctiveness and stresses an 'insider view'. Great touristic attractiveness, a complicated history and the fact of being an island (one of a few in Poland) are among the most important reasons for focusing on this area. Owing to close connections of Wolin with tourism functions, tourism must receive special attention in the study. The author treats identity as a complex phenomenon, referring both to its spatial and cultural meanings; however, the socio-cultural dimension is regarded as an overriding concern, which has been complemented by spatial characteristics.

The multidimensional character of the study required the application of a combined method of research. It included analyses of relevant scientific literature, spatial policy of the region, press materials and tourist information. Conclusions were also drawn from interviews conducted in 2011 and earlier (e.g. in 2001 and 2005). The most recent interviews with members of local communities, which provided output of close relevance to the theme explored in the current study, were focused on tourism and perception of the area. They consisted mostly of open-ended questions and embraced a group of 52 respondents. Regarding respondents' structure, a slight predominance of women (28) occurred; age distribution shows a larger share of mature people (especially those between 21–30 and 41–50 years old) and having secondary education. These interviews were augmented in 2011 by meetings with several authorities engaged in tourism development of the Wolin Island (Bożętka, 2011). The analysis has also integrated the results of two other surveys. The first examined regional identity of inhabitants and was provided by the Western Pomerania Province (Tożsamość regionalna..., 2011), the second investigated perception and attitudes to the Wolin National Park and was conducted by the Park in 2010 (Woliński Park Narodowy...., 2010). A set of observations carried out by the author in the period 1993–2011 supplements an empirical part of the research.

Since conceptions of identity play a principal role in the work, a review of the evolution of theories of identity became a significant part of the study.

IDENTITY AND REGIONAL IDENTITY-THEORETICAL FOUNDATIONS

Given the focus on the issue of island identity, in this case on Wolin, insight into the term "identity" is necessary. Identity can be explained from a psycho-cultural perspective that emphasises the individual, the group or even a culture, but it can also be regarded using a chorological approach, which is characteristic of sciences engaged in examination of spatial phenomena. The latter perspective is shared by geography; in this field of knowledge, identity is usually closely connected with a given level of a spatial scale, and as a result, concepts of local identity and regional identity have emerged. However, investigation into such a complex issue as identity requires a deeper understanding than a single discipline can provide. The author thus combines the two aforementioned positions.

Although many different definitions can be found, both sociology and psychology regard identity as the sense of self (Dictionary of Sociology, 2001 and The Corsini Encyclopedia of Psychology, 2010). Barney Warf in Encyclopedia of Human Geography (2006: 239) links both approaches, stating that "identity is simultaneously a deeply personal phenomenon and a social phenomenon that reflects, and in turn shapes, individual and collective behaviour". According to him, there are specific characteristics of the phenomenon. Firstly, identities both constitute and are constituted by the social world. Secondly, they are always historically specific and thirdly, they are simultaneously space forming and space formed ("Space affects not only what we see in the world, but also how we see it", Warf, 2006: 240). These preconditions play a significant role in the present study. Interestingly, human sciences usually link identity with man, treated individually or as a group, but geography, architecture and other space-related disciplines show that identity may also accompany any spatial unit, e.g. place, region or landscape.

Though being one of the most common forms of identity, regional identity is a very problematic construction and quite a difficult term to define. An important conceptual distinction lies between "identity of the region" and "regional identity of its inhabitants". Identity of a region can be divided into two categories: Images, which can also be conceived as parts of regional consciousness and which have a rather subjective character, and more "objective" classifications based on physical environment, culture, landscape, etc. (Paasi, 1986, Paasi, 2003 and Knapp, 2003). Ipsen (1997) identifies regional phenomena processed by individual perception, which are of considerable importance for forming a feeling of belonging. These qualities originate from place properties and comprise the contour, complexity, and coherence.

As far as regional identity is concerned, notions that traditionally accompany the idea of place – such as a sense of belonging to and an attachment to a given area – play a crucial role. The relationship between identity and bonds between people and places (areas) constitute a significant theoretical issue. It seems that identity can exist even without considerable links between people and a given piece of land. Nevertheless, their presence and their strength contribute greatly to the content and values of identity. Norberg- Schulz (2000) when highlighting connections between place and region stated that identifying with a place or an area largely contributes to regional identity, since identity is closely related to the feeling of place.

Taking the above into account, two principal aspects of regional identity can be distinguished:

1.Socio-cultural, focusing on people, especially inhabitants, their feelings and perceptions as related to social and cultural systems, and adequate elements of these systems

2.Spatial, centred on characteristics of a given area, sometimes close to the concept of "landscape identity".

Perhaps it is worth noting that the latter aspect cannot omit human activity and that the two perspectives co-exist, demonstrating important interrelationships.

TOURISM AND IDENTITY

Looking at identity through the lens of tourism, positive and negative influences of tourism can be found, both of which are intimately connected with cultural effects. Tourism can bring considerable benefits by enriching local or regional culture with new modes of behaviour, habits, and customs. It usually introduces (and simultaneously requires) an attitude of tolerance, broadens horizons of inhabitants (Przecławski, 1994), and intensifies cultural exchange. Furthermore, tourism in many ways emphasises the idea of culture as a negotiated and mutable phenomenon.

However, the growing impact of tourism may constitute a serious threat to local societies. This increasing pressure is widely explored in the literature. Tourism can be seen as an economic opportunity and a tool for local development as reported by many case studies (e.g. Marjavaara, 2007), but apart from this, it may lead to social exclusion and even contribute to decline of many destinations (see Agarwal and Brunt (2006) who demonstrate changes within English seaside resorts). Changes in local and regional cultures have been analysed not only empirically (e.g. MacLeod, 2004), but also from theoretical and philosophical points of view (e.g. Murphy, 1985, Hollinshead, 1993, Przecławski, 1994, Urry, 1994, Edensor, 2001, Hunter, 2001 and Wang, 1999).

Fundamental concerns involve treating culture as a commodity (e.g. Hunter, 2001), the "demonstration effect" (Murphy, 1985) connected with acculturation to the dominant interest (Nuňez, 1989 and Hunter, 2001), an increasing tendency toward simulation and the proliferation of images (Urry, 1994). As Palmer (1999) has indicated, when images are marketed and sold they contribute enormously to the rapid growth of heritage tourism. The force of touristic images is very strong and it affects local people, who may be identified and encountered only through constructed images. Noteworthy, heritage tourism often employs re-constructed or created landscapes. Many researchers argue that landscapes and their images can construct national identity (e.g. Gruffudd et al., 1991).

Importantly, expansion of cultural contact can be seen as an invitation to cultural hybridity. As Hunter (2001) contends, tourism accelerates change and reinforces the subsequent cultural effects. Processes such as construction of heritage and manipulation of ideology prove the strength of these impacts.

Though it is not the intention here to enter a debate about theories of cultural change in the contemporary world, some general theses should be invoked. A shift from modernity, postulated to involve "opening out to others" (Giddens, 1991 and Urry, 1994) to post modernity, with its specific and powerful desire for simulations (Wang, 1999, Hunter, 2001 and Edensor, 2001) connected with consumption of images (Urry, 1994) brings about a cumulative effect that deeply influences culture and identity. Hence, tourism can be regarded as a sphere of influence "where cultures overlap and fractalise" (Hunter, 2001).

Of course, the impact of tourism on identity goes beyond a strictly cultural sphere – with high complexity of material representations it strongly influences a physical layer of a landscape and spatial use. In general, destination areas frequently experience growing congestion, the homogenisation of architecture and land-use patterns. Heavy consumption of space that is useful for investment is accompanied by the tendency to aestheticise everyday life and by the popularisation of "themed environments" (Urry, 1994) displaying simulated realities. Consequently, in addition to producing areas deprived of genuine character, tourism may encourage a mental detachment from the land. The human dimension of place ('place' seen in different scales) reflected in the presence of bonds between humans and their environment encapsulated in the Yi-fu Tuan's concept of Topophilia (1974) may be evoked here. Affective ties with material environment differ in intensity, subtlety, and mode of expression (Tuan, 1974: 93) and are extremely important for forming the state of identification with a place, a feeling of belonging to and having roots in an exact place.

THE WOLIN ISLAND

Geographic attributes

The island of Wolin is an island surrounded by an inland lagoon, the open sea and rivers. Varied landforms, the influence of the sea, rivers and lakes, and a mosaic of different ecosystems form a unique natural landscape. The island stretches along the southern coastline of the Baltic Sea, in north-western Poland (Fig. 1). Poland has fewer than 50 islands, of which Wolin, at 250 km^2 and with a population of about 30,000 – is the largest. Characteristically, as mentioned before, Wolin is situated among three different water types – The Pomorska Gulf, a part of the open sea extending to the north, The Szczecin Lagoon (Zalew Szczeciński) in the west, and the rivers of Oder and its tributaries (Świna and Dziwna) in the south and east (Fig. 1).

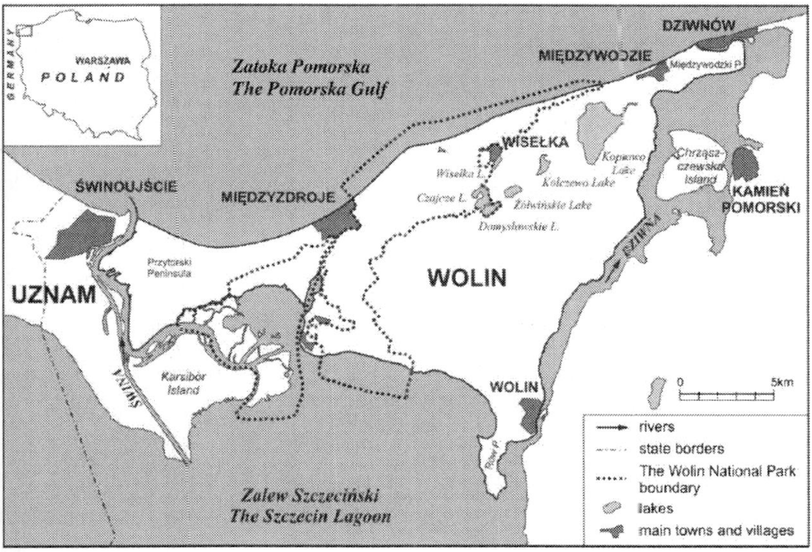

Figure 1. Wolin (source: the author's own work).

Wolin is adjacent to the island of Uznam, which in majority belongs to Germany. These two islands lie close to the Polish-German border and to the mainland. Wolin is a part of the Western Pomerania Province (województwo zachodniopomorskie) and a district of Kamień Pomorski.

Its territory is shared by four communities: Świnoujście, Międzyzdroje, Wolin and Dziwnów.

The island is featured by varied relief and different landforms. Key geomorphologic processes: aeolian and coastal form dunes, sandy beaches and high cliffs (Kostrzewski and Stryjakiewicz, 1986). The climate is temperate, with many oceanic features. This is the mildest climate in Poland and demonstrates high summer insolation comparing to the rest of the country. Important balneological values (thermal baths) occur locally, for instance, in Międzyzdroje (Młodzikowski, 1986 and Nowacki, 2000). The island exhibits rich biodiversity: 68 fish species, more than 230 bird species and 30 species of mammals including a unique European bison (Piotrowska et al., 2000 and Bereszyński et al., 2000). The role of the Wolin National Park, which protects biota and landscape and embraces a rich mosaic of habitats featuring different ecological character, with forest communities, dune plants and wetlands should be highlighted. Established in 1960 (4844 ha) and enlarged in 1966 to include coastal waters of the Baltic Sea and flood waters of the Świna's delta (10,937 ha) became the first (one of two) marine national park in the country.

The impressive landscape pattern found in Wolin- of a long beach accompanied by cliffs covered with lush forest greenery (Fig. 2), the pattern, which exposes constant activity of the sea, with its natural beauty and changeability, may be regarded as an inestimable value and simultaneously, an essential constituent of the island's identity.

Historical attributes

Wolin was well known during the Roman period thanks to amber trade routes and settlements of Goths. The ancient town of Wolin, in the south of the island, had a special reputation of being one of the richest ports in this part of Europe and the most beautiful harbour in the Baltic Sea. In the 10th century the island was incorporated into the Polish country formed by the Piasts and was granted a large degree of independence. Since the 12th century, it was troubled with many political conflicts that involved Denmark, Sweden and Germany. In 1762, Wolin became a part of Prussia and belonged to Germany until 1945. During Second World War, the island suffered damage, and after the war, it was returned to Poland as a part of the Reclaimed Land (Northern and Western Territories).

Figure 2. High moraine cliffs in the north of the Wolin Island are covered with beech and pine forests. The cliffs have recently been affected by strong abrasion processes (photo by author).

As a result of this turbulent history, only remains of tangible cultural heritage can be found, e.g. fragments of ancient settlements (including a reconstructed "town of Vikings" in Wolin); churches in Ładzin, Przytór, Międzyzdroje and Wolin; and a fortress in Świnoujście. Together with remnants of old Slavonic settlements, old fishermen's houses (e.g. in Wapnica), a draw- bridge in Dziwnów, lighthouses in Wisełka and Świnoujście, and rocket launchers in Wicko, these objects constitute the most important historical attractions. However, the spa areas in Międzyzdroje and Świnoujście are of special importance for tourism. Additionally, several cultural events boost the area's attractiveness: Jomsborg Wolin Festival of Vikings and Slavonic Nations (Wolin), International Choir Festival (Międzyzdroje), International Music Festival "Amber Baltic" (Międzyzdroje) and Holiday Artistic Festival (Międzyzdroje).

The island's economy has been based on harbours and sea transport (Świnoujście, Dziwnów), fishery, agriculture (Wolin commune), forestry, and tourism. Owing to its great assets, nature conservation plays a significant role, but this is tourism that tends to dominate all other activities.

TOURISM ON WOLIN

Thanks to its extraordinary natural values as a seaside location, the beautiful landscape, favourable local climate, and established tourism traditions, Wolin is a very popular destination among Polish visitors and holiday-makers. Its attractiveness also results from the presence of a wide range of touristic facilities (Tomczyk, 2004). Tourism here follows the rule of "Sea, Sand and Sun" and shows a typical mass tourism character, particularly in the northern part of the island during the summer season (June–August). In contrast, central and southern parts of the island feature extensive rather than intensive use. Touristic traditions are linked with development of two major towns: Międzyzdroje and Świnoujście, which were popular holiday resorts before the Second World War (Figs. 3 and 4); their spa functions are reflected in the name of Międzyzdroje (which means "between spas"). Today, the majority of visitors to the island come from Poland, particularly from the regions of Wielkopolska, Silesia, and nearby Western Pomerania. However, the share of foreign visits is not modest, mainly owing to arrivals from Germany and Sweden, and in a large degree because of the development of health tourism.

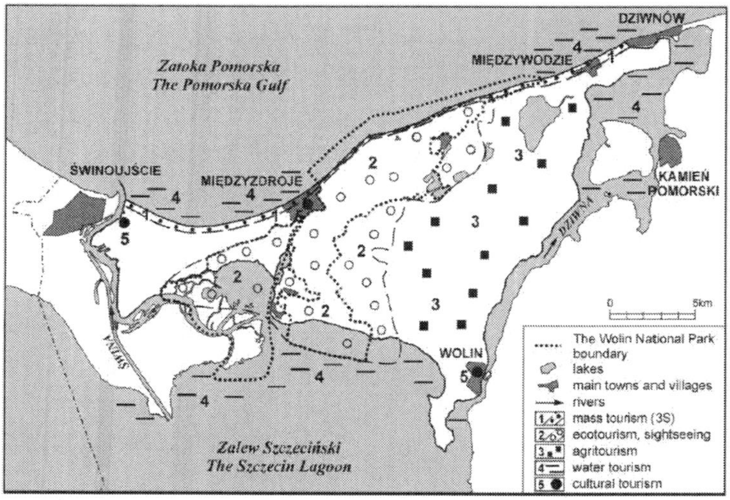

Figure 3. Distribution of main kinds of tourism within the Wolin island (source: author's own work).

Figure 4. Międzyzdroje (Misdroy in German) was a popular tourist destination before the war (Fig. 4). The pier is currently among the longest piers in Poland and offers a beautiful view on the north-western part of the island (Fig. 5, photo by author).

Międzyzdroje performs the role of the island's major tourist centre and a beach resort. It is also a functional microregion (Budner and Kaczmarek, 1986 and Sołowiej, 1992) of the north-western part of the island. As Sołowiej (1992) indicated, the town exhibits specific cycles of tourism activities – a dominant cycle involves beach tourism in the summer and is complemented by spa activities after the peak season.

Wolin has great natural assets for active tourism, especially water sports (sailing and canoeing) and cycling. However, these favourable preconditions are rarely exploited. In addition to beach activities tourists prefer sightseeing, mainly in relations to the Wolin National Park's. In terms of annual number of visitors, the Park usually ranks as the second or third most popular national park in the country (Jakuczun, 2004). Official data (Environment, 2012) recorded 1.5 million visitors in 2011. This protected area provides plenty of opportunity for land- and water-based ecotourism. The eastern and central parts of the island, demonstrating more of an agricultural character, are attempting to develop agritourism. Cultural tourism plays a less important role, but several events, such as Jomsborg Wolin Festival and Holiday Artistic Festival in Międzyzdroje have become very popular regional attractions. The town of Wolin should be taken into account owing to interesting exhibitions of its early medieval history accompanied by numerous cultural events focusing on traditions of Vikings and Slavonic people.

Interestingly, tourism on the island shows a specific pattern of distribution (Fig. 3). Cultural tourism is limited to major towns and villages and demonstrates a high degree of concentration. Other forms of tourism cover large areas and occupy particular zones. A network of touristic trails links the island's main attractions to one another and to waterways of the Szczecin Lagoon.

POST-WAR HISTORY OF WOLIN AND IDENTITY ISSUES

As previously mentioned, during its history Wolin was under various political influences, including the rules of Poland, Sweden, Denmark, and Germany, resulting in waves of cultural change. From 1762 to 1945, the island belonged to Germany. After the Second World War, as a part of the Northern and Western Territories, the area was returned to Poland. This change of state borders was connected with significant migration and resettlement processes in the 1940s and 1950s. German inhabitants left (about 9 million within the Northern and Western

Territories, including 4 million from Western Pomerania), and new settlers arrived (Chmielewski, 1998 and Wysiedlenia and i ucieczki 1939–1945, 2008). At first, immigrants to Western Pomerania arrived from neighbouring provinces on their own, but later the migration was strictly controlled by the state. The majority of settlers was of Polish origin and came from overcrowded provinces of central Poland or from previous Polish eastern territories, which were incorporated into the Soviet Union during the war (nowadays constituting parts of Byelorussia, Lithuania and Ukraine). A minority (fewer than 2%) arrived from Western Europe (Chmielewski, 1998). Additionally, Western Pomerania was a destination for numerous ethnic minorities, mainly Ukrainian and German. In consequence, this province is reported to have one of the most complicated socio-cultural structures in Poland (Chmielewski, 1998, Machaj, 2005 and Wysiedlenia and i ucieczki 1939–1945, 2008).

The results of the above changes are very serious. Firstly, cultural continuity was disrupted, since the new settlers represented other nations, cultures, and religions. Secondly, newcomers had many problems adjusting to a new land, where they encountered traditions, patterns of land-use, and architecture that were not appreciated. In fact, most of the immigrants did not accept the new situation and this reluctant attitude was continuously turning into alienation and indifference. The state of detachment from the land has two aspects: feelings of detachment from the 'core' Polish territory, and a lack of acceptance of the new area.[1] Groups of settlers differed culturally and large distances were established between the different groups (Machaj, 2005). It is worth noting that the new post-war communist state inhibited processes of autonomous social and cultural integration by preventing the development of local societies and identities. Reconstruction of former social and cultural links was not encouraged and a common law that allowed the area to be governed by local communities was suspended (Pawłowska, 1996 and Trosiak, 2008). Local and regional identities were almost abandoned concepts throughout the country during the post-war decades and became a matter of special scientific insight only after the political turnover in 1989 (Trosiak, 2008).

In consequence, local societies have been weak, cultural alienation has been experienced and the Northern and Western Territories have remained a kind of a foreign country for years. Weak identities or even a lack of regional identities of the inhabitants became an essential problem in the "new territories". Along with cultural uncertainty and a lack of cohesion, important social problems have appeared. According to official reports, Western Pomerania has been the province of Poland most afflicted by crime and social pathology right up to the present day (Strategia rozwoju..., 2010). Additionally, a passive attitude to social marginalisation and low levels of local activity have been noted (ibidem). A recent survey of regional identity of Western Pomerania inhabitants (Tożsamość regionalna..., 2011) revealed that more than one third of residents (37.2%) declared a lack of attachment to any place in the province.

Fortunately, as far as social and economic problems are concerned, Wolin's situation has not been as difficult as that of other parts of Western Pomerania. The reasons lie in possibilities opened up by tourism. Settlers to Wolin found relatively favourable economic conditions, and the area did not suffer severe physical destruction during the war. Throughout the post-war decades and today, the island has benefited from relatively high economic growth (Plan zagospodarowania..., 2002), a wider range of cultural attractions and less depopulation as compared with other areas in the province (e.g. Plan zagospodarowania przestrzennego województwa zachodniopomorskiego, 2002 and Studium uwarunkowań i kierunków zagospodarowania przestrzennego gminy Międzyzdroje, 2010a).

TOURISM AND THE SOCIO-CULTURAL DIMENSION OF WOLIN'S IDENTITY

As highlighted before, the processes of habituation of immigrants to new conditions on Wolin were ameliorated by the benefits offered by tourism opportunities. However, this could not completely eradicate feelings of detachment and alienation.

An analysis of regional planning policy confirms an influential role of tourism for development in the province (e.g. Plan zagospodarowania przestrzennego województwa zachodniopomorskiego, 2002 and Plan zagospodarowania przestrzennego województwa zachodniopomorskiego, 2010), particularly in coastal areas, including Wolin Island (Plan Rozwoju Lokalnego Powiatu Kamieńskiego na lata 2007–2013, 2006, Plan Rozwoju Lokalnego Gminy Wolin do 2013 r, 2007, Strategia rozwoju miasta i gminy Międzyzdroje na lata 2004–

2013, 2004, Studium uwarunkowań i kierunków zagospodarowania przestrzennego gminy Międzyzdroje, 2010a and Studium uwarunkowań i kierunków zagospodarowania przestrzennego gminy Wolin, 2010b). Taking into account the role of tourism in economic development and its strong impacts on culture, tourism's role in identity formation requires closer attention.

After the political turnover in 1989 a need to build and strengthen local and regional identity has become one of the most important aims for local governments. Enhancement of identity and socio-cultural integration is among the principal tasks designated for Western Pomerania for the coming years (Strategia rozwoju..., 2010). The most important concerns comprise an increase in social cohesion, and the development of cultural heritage and growth of a sense of cultural belonging, which are considered to form an essential basis for regional identity (Strategia rozwoju..., 2010: 111).

Numerous newspaper articles, formal analyses and discussions convey that the state of "looking for identity" has become prevalent in socio-cultural life. Remarkably, the search for identity is connected with a search for meaning. Elements that could help build associations and act as distinguishing marks are much needed, and in fact, almost everything is deemed helpful if it can arouse feelings of like and pride. This at least partially explains the hunger for new attractions and the really strong desire demonstrated by local people to encounter and host tourists. Attractions, events, and a feeling, that "something is taking place" has become a real need expressed by the people.[2]

The popularity of tourism stems in part from its provision of attractions and cultural goods. Importantly, a hospitable attitude to tourism is characteristic of the island's communities. Sociological research on tourism development on Wolin has demonstrated a characteristic affirmation for tourism among local citizens and a willingness to invite guests (e.g. Łyczkowska, 1995 and Matuszewska, 2003), though some alteration in the inhabitants' perception of tourism has also been observed (Kostrzewski and Stryjakiewicz, 1986 and Ewertowski, 2003). The research conducted by the author shows that the island is associated by its inhabitants mainly with tourism (94.1% of respondents declared tourism as one of the most important associations of Wolin).

Subsequently, it can be concluded that opportunities provided on Wolin by touristic traditions established before the war (Fig. 4) and because of post-war development of the sector were not limited to economic benefits. It may be assumed that the greater role of tourism was to enhance cultural exchanges and open the island to domestic and foreign arrivals. Importantly, the increase in arrivals of people from different parts of Poland helped create stronger bonds between the local population and their area than was evident in more remote corners of the "new territories", and it also enabled the island to gradually lose its peripheral status. Tourism made possible to become closer to the 'outside world' and this world was viewed through encounters with tourists. This process may be understood from a psychological perspective – the outside world is seen through the encounters with visitors (seems to be 'at hand'), and using a more concrete, physical perspective – real distances have shortened thanks to transport routes connecting the island and the rest of the country. Interestingly, only two persons examined in the last author's research complained about isolation or a kind of territorial disconnection that frequently accompanies islands.

Tourism can emphasise local character and activate expansion of local culture, and indeed, most cultural events on the island are tourist – oriented. However, Wolin lacks unique cultural patterns, and the roots of local communities or their specific ethnic cultures are not exhibited. Although cultural events are very popular and are eventually associated with particular places, they do not demonstrate the special local values that help constitute *genius loci* or a sense of place. Rather, the events eagerly use seaside conditions and beach resorts assets, in a common and general sense, without relevance to precise places.

With one exception, the island's material traces of history do not exhibit exceptional values. The authority of the town of Wolin has implemented an ambitious project based on archeological findings and the historic reconstruction "Wolin- a Town of History" (Plan Rozwoju Lokalnego..., 2007). The project explores the early Middle Ages, an epoch of Vikings and ancient Slavonic tribes, and is composed of several elements, including the 'Centre of Vikings and Slavonic People open-air archeological museum', the 'Niedamira historic shipyard' and the 'Jomsborg Wolin Festival of Vikings and Slavonic Nations'. The undertaking results in the reconstruction of a compact medieval section in the town of Wolin. The nearby historic shipyard on the Dziwna River aims to build replicas of old ships and boats and to promote Baltic boat-building traditions. One of the points is that medieval Wolin is identified in Icelandic sagas with the legendary Jomsborg, a defensive settlement of Scandinavian and Slavonic Vikings. The Jomsborg Wolin Festival has become a spectacular event, with about 800 exhibitors and tens of thousands of tourists. It lasts 3 days and demonstrates ancient Viking and Slavonic life, including their craft, armour and social ceremonies. Participants spend nights in tents, get involved in craft workshops, try to build goods out of purple willow, construct bows, fight, prepare Viking food, etc. The scenery is complemented by characteristic silhouettes of old ships smoothly sailing the river and lagoon.

However, it is surprising that along with this ambitious initiative, the commune organises another attraction in the form of a Western City theme park located in the small village of Sułomino near the town of Wolin.[3]

The developing "Wolin- a Town of History" project fuels the imaginations of tourists and local people. No doubt, it exploits the island's heritage and presents a simulation. Such advanced operations of the heritage industry necessarily affect identity. As it is acknowledged, relationship among tourism, identity (exceptionally, place and political identity) and heritage resources are very close (e.g. Ashworth, 1995). According to Palmer (1999: 318), museums, theme parks and related attractions can provide an opportunity to reaffirm a sense of belonging. However, managing the imagination can involve the manipulation of social awareness, making the effects of heritage tourism on identity ambiguous. To what extent does the enterprise conducted in the Wolin commune influence regional identity? The answer is not simple. The project may strengthen people's links with the island by highlighting its interesting history and it can promote the island in Scandinavia, but it may be very difficult for contemporary citizens to identify with Vikings. Although these events and places create the sense of uniqueness, this is only partly rooted in the culture of today's inhabitants (fortunately, though not always focusing on it, the event exhibits some traditions of Slavonic tribes). Nevertheless, a key function of the project consists in creation of attractions and fulfillment of demands expressed by locals as well as tourists. Local residents view the enterprise as a positive impact on the island's attractiveness (Bożętka, 2011). Importantly, Wolin's interest in the Viking heritage is not connected with national preferences, which for example characterises Shetland Islands (Grydehøj, 2008).

The process of heritagisation usually raises many questions, which tend to focus on the degree of authenticity, the relevance of the past to the present and manipulation of images of the past. It is asserted that islands are especially inclined to be under influence of heritage tourism (e.g. Ronstrőm, 2008) and some of them experience enormous production of pasts. Ronstrőm (ibidem), discussing a case of Gotland shows that heritage became not only a special interest, but also a major industry in this island.[4]

The influence of tourism on the island's identity must be regarded in terms of the impact of foreign cultures. The open attitude to tourism held by Wolin's inhabitants is followed by an easy reception for patterns of behaviour and cultural impacts that come from abroad. The influences embrace such typical cultural activities as dance and music, displaying a widespread tendency to employ international pop and rarely demonstrate more ambitious inclinations, for instance to ethno music. Foreign influences also affect gastronomy, as numerous national cuisines compete successfully with local seafood.

Interestingly, the habit of celebrating important dates commemorating war and particularly, the post-war period of 'regaining' the territory, which used to be highly respected by the communist state, has been replaced by an enormous popularity of new festivals typical of Western countries, such as Valentine's Day and St. Patrick's Day. According to the author's study (Bożętka, 2011), majority of respondents appreciated the presence of new cultural forms, especially these connected with music and gastronomy, however, some interviewees considered the changes to be rather odd and their rate to be too fast.

Different cultural patterns are mixed and frequently resemble a state of chaos, but foreign influences do not yet dominate core elements of Polish culture. However, the predominance of German visitors among foreign arrivals, and the island's closeness to Germany leads to this national group having major impacts, both on tourism and on the economy of the island. The popularity of the German language grows alongside the increase in a number of German tourists. Nowadays most cultural information, including descriptions of landmarks and menus in restaurants is bilingual. Preferences of this nation play a great role in tourism management on the island. The prices of many goods, including real estate, are directed at this market segment, and such a process is not appreciated by Polish visitors. German influences are not, however, caused only by tourism, because many people from Western Pomerania and from Wolin itself work in the neighbouring country and even live there with their families. Nonetheless, in light of the history of the island, this impact may be considered to have serious consequences on identity and on the sociocultural sphere, not excluding emergence of hazards for national self-recognition.

Influences from abroad can be observed in the changes leading to adoption of less traditional ways of life, whereas affirmation of traditional values seems to be characteristic of Poland. Besides that, it must be noted that this sort of changes usually results from the operation of many factors, and frequently occurs in tourist areas.

SPATIAL DIMENSIONS OF IDENTITY

Tourism overwhelmingly influences spatial aspects of identity. This theme constitutes a very contentious problem that requires further elaboration, which exceeds the scope of the current study. Nevertheless, the following section will focus on the contribution of spatial elements to socio-cultural identity.

The impacts of a tourist function on spatial use and the landscape of Wolin are strong and widespread. The growth of tourist areas has a dynamic character, but the rate of change is not assisted by efforts to maintain adequate landscape values. In fact, a loss of the visual quality and individuality of the island's spa resorts has been characteristic of the post-war period (Bal, 2009). Deterioration in the form and in the content of sea resorts involves both their general architectural idea and aesthetic qualities. Chaotic and eclectic character of architecture is accompanied by a loss of spatial coherence and connections with nature (Bal, 2002 and Bal, 2009). The great popularity of new, multicultural patterns marginalises old, historic objects and invites a cosmopolitan scheme of "anywhere places". Sadly, acceleration of these changes can now be observed in Międzyzdroje (Fig. 5). For instance, the famous "Amber Hotel", a symbol of spatial and economic transformation in the beginning of 1990s, even considered to be a landmark is now surrounded by many other similar uniform hotels.

Figure 5. A growth in tourism leads to disturbance of composition of built-up areas. Contrasts between old residences and new apartment blocks in the spa of Międzyzdroje cause visual tension (photo: author).

Tourism threatens also outstanding natural landscape values, not excluding the area protected by the Wolin National Park. Several zones, for instance in the vicinities of Międzyzdroje and Wisełka have been facing serious conflict between tourism and nature protection for decades (Sołowiej, 1992, Bożętka, 1995 and Macias et al., 1995). Noteworthy, along with the pressure for economic growth the role of natural values decreases in social awareness. Tourism can thus be seen as a source of physical degradation of a natural landscape and furthermore, of its mental devaluation. Obviously, this poses a danger for nature and the landscape, but it is also harmful for the island's socio-cultural identity. The great capacity of the island's nature and landscape for creating or strengthening identity is underestimated. It should be noted that none of the examined spatial documents and formal plans for development in the region indicated the role of nature in the self-recognition of the inhabitants, even though problems of identity are frequently stressed.

On the other hand, tourism can contribute to enhancement of place identity-this process is observed in the town of Wolin's efforts to emphasize heritage. Tourism in this case serves as a crucial factor leading to a far-reaching transformation of the mental landscape and leading to the occurrence of a new element that extensively influences the island's image.

The image can exist far away from their origins, being encapsulated in visitors' minds. Regrettably, Wolin lacks research on its image, but some information can be found indirectly, in analyses devoted to touristic attractiveness of the island. The works of Sołowiej, 1992 and Łyczkowska, 1995 distinguish the beach, The Wolin National Park, and its cliffs and animals (especially bison), which perform as principal elements of the island's image. Matuszewska (1995) investigating perception of the national park observed poor recognition of the area and its boundaries. She revealed that many respondents were not aware that they spent their holidays on an island.

Islands, as Hay argues (2006) may be regarded as special places. Meanings, which transform islands into place may be more pronounced, better articulated, and more effectively defended than in other areas (p. 34). Noteworthy, the metaphorical idea of an island possesses a great power (see e.g. Baldacchino, 2008), and the role of landscape for creating image and identity of the island may be enormous. The Caribbean provide a particular instance of a strength of the associations built upon a vision of a tropical landscape – the visual landscape has long been a central focal point in narratives about these islands and instigated a rise of tourism (Nelson, 2011). We can expect that the emphasis in islands' images is put on an island. However, Wolin does not completely fit in this thesis. The analysis of official statements and plans, including incentives for tourism development, shows that the fact of being an island does not play an important role, it is overshadowed by general assets of location by the sea.[5] That sounds quite surprising regarding islands' advantages tourism industry frequently and eagerly exploits. It seems that one of few representatives of the idea of islandness is weekly Wyspiarz Niebieski ('The Blue Islander', 2006-2011), which puts some energy into highlighting a special character of Wolin, Uznam and adjacent islets. The magazine reports and advocates elections of 'The Islander of the Year', frequently mentions deficiency of sea issues in regional policy and sometimes publishes recollections of citizens regarding 'a life on 44 islands'.

EXTERNAL SITUATION AND THREATS FOR IDENTITY

Identity is a very subtle phenomenon, absorbing a multitude of influences. It can be both fragile and persistent. Furthermore, it depends not only on site-specific factors, but also on worldwide processes. The impact of interrelated multi-scale processes can bring many threats to numerous areas, especially those which are isolated, underdeveloped, and faced with strong external influences. Certainly, this is the case of many islands. As far as Wolin Island is concerned, the unstable and immature character of its identity may cause problems, leading to increasingly negative impacts in the future.

Enhancement of identity and strengthening the links between people and places in an epoch of strong globalisation and influential multicultural patterns do constitute challengous tasks. Two contemporary tendencies can extensively impact Wolin's identity. One of them is connected with intense emigration of Polish citizens to Western Europe since Poland's accession to the European Union in 2004. Although Wolin's rate of depopulation is slower than that of other parts of Western Pomerania, emigration abroad, especially of young people, is taking place. Another factor is the global cultural tendency toward multiculturality.

The strength of multicultural influences on Wolin is noticeable, though perhaps not so unbearable as in many other parts of the world. Importantly, multiculturality does not only affect the community level. Adler (2007) poses a question concerning a new type of personality, contrasting the multicultural individual with the culture-bound individual. According to his research, the identity of a multicultural person is more fluid and mobile, more open to variations. It is based on a cultural context rather than on "belonging to". A multicultural person recreates his identity, changes easily to fit the context, exhibits great mental flexibility and adaptability. Adjustment replaces belonging and traditional values. The multicultural individual's world lacks coherence: he or she seems to live on the boundaries of societies. This convincing description provides only a small modicum of hope for those who regret the disappearing links-with culture, landscape, and a piece of land.

Negative influences on the island's identity are also caused by the continuous loss of spatial identity in Poland in the post-war period. During the socio-economic transformation after 1989, new forms of landscape degradation were set in motion, resulting for instance from attempts to privatise public space. An increase of conflicts around the Wolin National Park and efforts to devaluate its mission illustrate a side-effect of economic growth.

Moreover, architecture and spatial planning in Poland has been in a serious crisis for years. As a consequence, uncontrolled settlement sprawl is accompanied by occurrence of numerous places, which are familiar to nobody and that are unable to evoke feelings of attachment and to form positive impacts on identity. The introduction of familiar patterns is particularly recommended in areas that, like Wolin, suffer problems with identity. Unfortunately, architecture in Poland rarely employs regionalistic and vernacular tendencies (e.g. Pawłowska, 1994, Pawłowska, 1996 and Basista, 2001). The above mentioned loss of identity in Polish towns and cities is widespread (Basista, 2001 and Bal, 2009) and rises serious threats to maintenance of heritage. Noteworthy, the problem of spatial identity is especially important, since the flow of homogeneity appears to be a main feature of post-modern landscapes (Antrop, 2005).

There is also one more danger for identity. It has a general character and comprises denying the role of identity, regardless the subject it involves: individuals, societies, regions, places or landscapes. Sadly, strong concentration on today, contemporary problems and economy can suppress more elusive notions such as memory, nature and identity.

THE EVOLVING CHARACTER OF IDENTITY

Identity may be considered an evolutionary process. It is constructed, contested and negotiated over long periods of time and is vulnerable to changes; with generational shifts being of high importance. New generations have grown up on Wolin Island without the experience of war or of dislocation from land. Though aware of their ancestor's history of migration, the younger age groups have become accustomed to the Wolin area. However, a lack of place-based roots and tradition lasts longer than the life-span of one generation and as a result, it affects the descendants. The social problems of Western Pomerania and the continuous state of "looking for" underline the importance of this problematic process.

Notably, the Northern and Western Territories are increasingly characterised by a process of substituting the search for identity with the search for material status (Machaj, 2005). Such a change may support the thesis that multiculturalism offers a new type of identity (Adler, 2007). As a consequence, there is a need to discuss the question of the potential deconstruction of the immature cultural identity of Wolin. Furthermore, general theory can prove that the opposite direction of evolution-devolution is also possible.

Two additional remarks should be expressed here. Firstly, although the fulfilment of economic requirements tends to play a leading role, this should not necessarily exclude other needs and values. Secondly, although identity is intimately connected with history, it is nevertheless a future-oriented process, exhibiting an open character and inviting changes and modifications.

CONCLUSIONS AND DISCUSSION

Issues of identity usually raise many discussion points. Identity is ascertained to build a value system and to be a value itself. However, narratives of identity involve a hidden hazard, which is instrumentalization, or the design of an instrument to meet ones goals. Identities demonstrate close connections with associations and images, are frequently embedded in heritage and its construction, and can be subordinated to governing powers. As Bauman (2004) indicates, social identities remain open, subject to negotiations and conflicts that are immersed in power relations. Noteworthy, any effort to construct identity, especially national identity, may be regarded as being opposed to a more desirable, autonomous and spontaneous process of identity formation. Essentially, from the perspective of ethics, adoption of an attitude towards identity that insists on control and development of steering mechanisms may lead to a highly awkward situation and may compromise the independence of individuals and groups.

Identity depends on many factors and cannot avoid being influenced by changing power structures. The case of Wolin illustrates how external changes can exert a great influence on evolution of identity. Transformation of the ethnic structures of Poland and the surrounding region, a shift of political systems and subsequently, of economic systems have significantly affected components of identity. Additionally, rapid changes in conditions in the 20th century were exacerbated by the steady influence of tourism on the life of the island. The strength of these influences can be seen in the dual nature of Wolin's identity: a pronounced spatial identity that can be stated in geo-physical terms is accompanied by a less definite socio-cultural aspects.

Tourism, as it was outlined above, accelerates change. It can therefore be anticipated that tourism development will lead to alterations in identity, some of them negative. Disadvantageous processes in Wolin may include the deterioration in natural values and a strong impact of German visitors on local economy and culture. However, the role of tourism on the island cannot be restricted to negative influences, rather, tourism can become a decisive factor in the creation or reinforcement of bonds between people and the area. A history of nostalgia for the Eastern Territories and a feeling of detachment gradually lose importance, as the attractiveness of the touristic region increases. Tourism, being both an external demand and a tool to meet expectations of local people, has a very strong and generally positive influence on public awareness of island identity. Acceptance of and a desire for the atmosphere and activities that tourism provides, predisposes this industry to play a vital role in constructing identity. The tourism-driven development of identity on Wolin has the potential to change thinking concerning the identity of the province, producing new aspects and positive values. More controversially, from a theoretical point of view, identity of the island can be regarded as a vast container absorbing a wide range of tendencies, attractions, and activities. In addition to that, one should discern potential dangers inherent in tourism's simulations and the manipulations of identity that are involved.

Importantly, Wolin does not demonstrate relations of power frequently found in islands, which consist in opposition of 'island-mainland', an impact of colonisation and problems of dependency (described for instance by Baldacchino, 2008). Its identity cannot be recognized only through the context of islandness. It is noteworthy that, in spite of being an island, Wolin is rarely associated with relevant descriptions. This is another week point not only in attempts to enhance identity, but also in undertakings connected with tourism development. Thus, there is rather a need to strengthen the awareness of being an island, bearing in mind that treating any territory as an island can enhance socio-cultural cohesion. Focus on an 'island' should also incorporate a concern of the role of the sea in constructing identities. Providing people on the mainland with a perception of Wolin as an island and introducing people the theme of sea and identity would bring a new dimension to the analysis and to the process of identity formation.

More specifically, it should be noted that the notion of islandness is frequently employed in the processes of heritagisation as in the case of Gotland (Ronstrőm, 2008). Again, Wolin does not completely adhere to the rule. Though construction of past observed in the town of Wolin involves the plot of an island, this does not constitute a key element of the enterprise; islandness accompanies 'contemporary Vikings and Slavonic people', but ancient ships would be built and battles would be fought even if the town had been located on the mainland (nevertheless, Wolin's islandness did have importance for genuine Vikings and ancient Slavonic people). Certainly, interrelations between identity and islandness are usually both interesting and significant, however, a loose link of Wolin with its own islandness shows that they are not unambiguous. This confirms Baldacchino's intuition, who writes that 'the state of islandness per se does not necessarily create the community basis for a unified island identity (Baldacchino, 2004, after Grydehøj, Hayward, 2011). Indistinct, weak or pale identities of islands may occur more frequently than it is expected- Grydehøj and Hayward (ibidem) in an analysis of the historical process of identity building of the Isle of Wight reveal considerable problems with recognition of the island's distinct character (the Isle of Wight is viewed in a very non-specific way, usually in the context of English countryside and seaside tradition).

Most of all, Wolin's identity cannot be compared to produced identities of many islands. An impact of tourism branding, heritage organisations, marketed and sold images has not been overwhelming here. Evolution of identity has depended in Wolin on more intrinsic features of societies and processes taking part in the state. Nonetheless, the situation may change with reorientation in economy and growing pressure of tourism industry. This is one of the reasons for making efforts to reinforce and highlight the island's identity, no matter what the stage of evolution of identity is. Maintenance of identity is essential in the context of a globalising world, some of islands have much experience in identity protection, for example Jersey, which implemented cultural strategy centred on its distinctive and unique character (Riddell, 2007).

In general, Wolin's identity depends on three main constructs: (1) natural features, (2) history and culture, (3) economy and land-use. However, these principal elements are shaped by more specific issues: post-war history and migration processes; tourism; islandness; nature and landscape. The following study reveals that these factors are featured by a varied intensity of influence. The impact of post-war history connected with migration processes, and tourism tends to play a leading role. This is noteworthy that island generally are strongly connected with migration processes (see e.g. King, 2009). Characteristically, a considerable discrepancy between socio-cultural and natural domains of the Wolin's identity constructs is found. Official documents on many occasions declare that cultural factors are of primary importance for the area's identity (even though if economic forces appear to be pre-eminent), but the role of natural assets is underestimated, despite the fact they hold great potential for creating identity. Particularly, the capacity of the Wolin National Park deserves more attention in formal analyses as well as in social discourse.

It seems that the cardinal question on the chances of the Wolin Island to retain its identity will be followed by the answer embracing maintenance and construction, while maintenance refers to Nature, construction to Culture.

ACKNOWLEDGEMENTS

I am deeply grateful to: Arkady, Elżbieta Karmowska, Anna Falkiewicz and Sylwia Łyczkowska for the inspiring work conducted in Wolin during several summer periods.

I thank very warmly Adam Grydehøj for reviewing and strengthening large portions of this study.

REFERENCES

1. Adler, P., 2007. Beyond cultural identity: reflections on multiculturalism.
2. In: Bennett, M.J. (Ed.), Basic Concepts of Intercultural Communication (selected readings). Intercultural Press, pp. 225– 275.
3. Agarwal, S., Brunt, P., 2006. Social exclusion and English seaside resorts. Tourism Management 27, 654–670.
4. Antrop, M., 2005. Why landscapes of the past are important for the future. Landscape and Urban Planning 70, 21–34.
5. Ashworth, G.J., 1995. Heritage, tourism and society: a European future for a European past? In: Herbert, D.T. (Ed.), Heritage, Tourism and Society. Mansell Publishing Limited, London.

6. Bal, W., 2002. Wpływ degradacji dziewie˛tnastowiecznych kurorto´w morskich na stan to_zsamos´ci kulturowej regionu nadmorskiego Pomorza Zachodniego. Architektura Krajobrazu 3–4, 122–127.
7. Bal, W., 2009. Cultural identity of seaside health resorts – an inspiring space and demanding heritage. Przestrzen´ i Forma 12 (2009), 143–149.
8. Baldacchino, G., 2008. Studying islands: on whose terms? Some epistemological and methodological challenges to the pursuit of island studies. Island Studies Journal 3 (1), 37–56.
9. Basista, A., 2001. Betonowe dziedzictwo, Wydawnictwo Naukowe PWN, Warszawa-Krako´ w, p. 296.
10. Bauman, Z., 2004. Identity: Conversations with Benedetto Vecchi.Polity Press, Cambridge.
11. Bereszyn´ ski, A., Dylawerski, M., Jakuczun, B., Jurczyszyn, M., Maciorowski, G., 2000. Stan zachowania fauny Wolin´ skiego Parku Narodowego oraz perspektywy jej ochrony i restytucji. Klify 4, 171–184.
12. Bo_ze˛tka, B., 1995. Antropopresja na obszarze Wolin´ skiego Parku Narodowego na tle konfiktowos´ci funkcji i problemu sa˛siedztwa. Klify 2, 51–54.
13. Bo_ze˛tka, B., 2011. Materials on Wolin, [documents, unpublished] Institute of Geography. Gdan´ sk University, Gdan´ sk.
14. Budner, W., Kaczmarek, T., 1986. 'Mikroregion funkcjonalny Mie˛dzyzdrojo´w i granice Wolin´ skiego Parku Narodowego a podział administracyjny wyspy Wolin', in: Kostrzewski, A. (Ed.), Wolin´ ski Park Narodowy. Monografia geograficzna, Studenckie Koło Naukowe Geografo´w im. St. Pawłowskiego, Uniwersytet im. A Mickiewicza w Poznaniu, Poznan´ , pp. 169–178.
15. Chmielewski, Z., 1998. Struktura ludnos´ciowa i narodowos´ciowa Pomorza Zachodniego po II wojnie s´wiatowe'j. in: Giedrojc´, M. Mieczkowski, J. (Eds.), Pomerania Ethnica. Mniejszos´ci narodowe i etniczne na Pomorzu Zachodnim, pp. 54–59.

16. Dictionary of Sociology, 2001. Lawson,T. Garrod, J. (Eds.), Dictionary of Sociology, Fitzroy Dearborn Publishers, London-Chicago p. 273.
17. Edensor, T., 2001. Performing tourism, staging tourism. (Re)producing tourist space and practice. Tourist Studies 1 (1), 59–81.
18. Environment, Statistical Yearbook 2012. Central Statistical Office of the Republic of Poland, Warsaw, p. 599.
19. Ewertowski, M., 2003. Mie˛dzyzdroje- kurort turystyczny zamiast miasta przyjaznego mieszkan´ com. In: Kostrzewski, A. (Ed.), Wolin´ ski Park Narodowy-s´rodowisko przyrodnicze, kształtowanie i ochrona. Bogucki Wydawnictwo Naukowe, Poznan´ , pp. 141–155.
20. Giddens, A., 1991. Modernity and Self-Identity. Self and Society in the Late Modern Age. Stanford University Press, p. 256.
21. Gruffudd, P., Daniels, S., Bishop, P., 1991. Guest editorial-landscape, heritage and national identity. Landscape Research 16 (2), 1–2.
22. Grydehøj, A., 2008. Branding from above: generic cultural branding in shetland and other islands. Island Studies Journal 3 (2), 175–198.
23. Grydehøj, A., Hayward, P., 2011. Autonomy Initiatives and Quintessential Englishness on the Isle of Wight. Island Studies Journal 6 (2), 179–202.
24. Hay, P., 2006. A phenomenology of islands. Island Studies Journal 1 (1), 19–42.
25. Hollinshead, K., 1993. Encounters in tourism. In: Khan, M.A., Olsen, M.D., Var, T. (Eds.), VNR's Encyclopedia of Hospitality and Tourism. Van Nostrand Rheinhold, NewYork, pp. 636–651.
26. Hunter, W.C., 2001. Trust between culture. The tourist. Current Issues in Tourism 4 (1), 42–67.
27. Ipsen, D., 1997. Raumbilder Kultur und }Okonomie ra¨ umlicher Entwicklung. Centaurus-Verlagsgesellschaft, Pfaffenweiler. Jakuczun, B., 2004. Turystyka w Wolin´ skim Parku Narodowym'.

in: Wyspa Wolin. Zro´wnowa_zony rozwo´ j w strefie przygranicznej, Oficyna In Plus, Szczecin, pp. 161–166.

28. King, R., 2009. Geography, Islands and Migration in an Era of Global Mobility. Island Studies Journal 4 (1), 53–84.

29. Knapp, W., 2003. Regional identity- a conceptual framework. in: <http://www.saulprojects.net> (accessed 03.01.2011.).

30. Kostrzewski, A., Stryjakiewicz, T., 1986. Funkcjonowanie Wolin´ skiego Parku Narodowego w systemie przyrodniczym i społeczno ekonomicznym wyspy Wolin'. in: Kostrzewski, A. (Ed.), Wolin´ ski Park Narodowy. Monografia geograficzna, Studenckie Koło Naukowe Geografo´w im. St. Pawłowskiego, Uniwersytet im. A Mickiewicza w Poznaniu, Poznan´ , pp. 9–16.

31. Łukasiewicz, K., 2002. Wie˛z´ , kto´ rej nikt zerwac´ nie zdoła. Magazyn Wilen´ ski 2 (2002), 24–25.

32. Łyczkowska, S., 1995.' System rekreacyjny miasta Mie˛dzyzdroje w s´wietle opinii turysto´w i mieszkan´ co´w miasta', Klify, vol. 2. pp. 97–111.

33. Macias, A., Karmowska, E., Bogajewski, T., 1995. Antropopresja na obszarze Wolin´ skiego Parku Narodowego i jego obrze_za pod wzgle˛dem ilos´ciowym i jakos´ciowym. Klify 2, 55–82.

34. Machaj, I., 2005. Społeczno-kulturowe konteksty to_zsamos´ci mieszkan´ co´w wschodniego i zachodniego pogranicza Polski. Wydawnictwo Naukowe Scholar, Warszawa, pp. 308.

35. MacLeod, D.V.L., 2004. Tourism, Globalisation and Cultural Change: an Island Community Perspective, Channel View Publications, p.244.

36. Marjavaara, R., 2007. Route to destruction? Second home tourism in small island communities. Island Studies Journal 2 (1), 27–46.

37. Matuszewska, D., 1995. Wolin´ ski Park Narodowy w s´wiadomos´ci turysto´ w. Klify 2, 113–135.

38. Matuszewska, D., 2003. Funkcje turystyczne i konflikty w wybranych parkach narodowych Polski po´ łnocno- zachodniej. Bogucki Wydawnictwo Naukowe, Poznan´ , 151.

39. Młodzikowski, Z., 1986. Klimat Wolin´ skiego Parku Narodowego. Wolin´ ski Park Narodowy. Monografia geograficzn. in: Kostrzewski, (Ed.), Studenckie Koło Naukowe Geografo´w im. St. Pawłowskiego, Uniwersytet im. A Mickiewicza w Poznaniu, oznan´ , pp. 24–31.
40. Murphy, P.E., 1985. Tourism. A Community Approach. Methuen, New York, p. 200.
41. Nelson, V., 2011. The landscape reputation: tourism and identity in the Caribbean. Journal of Economic and Social Geography 102 (2), 176–187.
42. Norberg-Schulz, C., 2000. Bycie, przestrzen´ i architektura. Wydawnictwo Murator, Warszawa, 120.
43. Nowacki, F., 2000. Ochrona ekosystemo´w wodnych Wolin´ skiego Parku Narodowego. Klify 4, 63–108.
44. Nunˇ ez, T.A., 1989. Touristic studies in anthropological perspective. In: Smith, V.L. (Ed.), Hosts and Guests: The Anthropology of Tourism. University of Pennsylvania Press, Philadelphia, pp. 265–279.
45. Paasi, A., 1986. The institutionalization of regions: a theoretical framework for the understanding of the emergence of regions and the constitition of regional identity. Fennia 164, 105–164.
46. Paasi, A., 2003. Region and place: regional identity in question. Progress in Human Geography 4, 475–485.
47. Palmer, C., 1999. Tourism and the symbols of identity. Tourism Management 1999, 313–321.
48. Pawłowska, K., 1994. O percepcji własnego miejsca i jej skutkach, czyli idei swojskos´ci architektury', Zeszyty Naukowe "Człowiek I S´ rodowisko" Polskiej Akademii Nauk, vol. 9. pp. 77–108.
49. Pawłowska, K., 1996. Idea swojskos´ci w urbanistyce i architekturze miejskiej, Monografia nr 203, seria Architektura. Politechnika Krakowska im. T, Kos´ciuszki, Krako´w. Piotrowska, H., Fałtynowicz, W., Markowski, R., Sagin, P., 2000. Wybrane problemy ochrony ekosystemo´w les´nych. Klify 4, 157– 164.

50. Plan Rozwoju Lokalnego Gminy Wolin do 2013 r, 2007. [Plan of the Local Development of the Wolin Commune till 2013], Wolin.

51. Plan Rozwoju Lokalnego Powiatu Kamien´ skiego na lata 2007–2013, 2006. [Plan of the local Development of the Kamien´ Pomorski District, 2001–2013], Kamien´ Pomorski.

52. Plan zagospodarowania przestrzennego wojewo´ dztwa zachodniopomorskiego, 2002, [Spatial Plan of the Zachodniopomorskie Province, 2002] Regionalne Biuro Gospodarki Przestrzennej, Szczecin.

53. Plan zagospodarowania przestrzennego wojewo´ dztwa zachodniopomorskiego, 2010. [Spatial Plan of Zachodniopomorskie Province, 2010], Regionalne Biuro Gospodarki Przestrzennej, Szczecin.

54. Przecławski, K., 1994. Tourism and Contemporary World. University of Warsaw, Warsaw, p. 100.

55. Riddell, A., 2007. Jersey: the development of an island cultural strategy, Shima, vol. 1. p. 1.

56. Ronstr}om, O., 2008. A different land. Heritage production in the island of Gotland, Shima, vol. 2. p. 2.

57. Sołowiej, D., 1992. Weryfikacja ocen integralnych atrakcyjnos´ci s´rodowiska przyrodniczego człowieka w wybranych systemach rekreacyjnych, Wydawnictwo Naukowe Uniwersytetu im. A Mickiew cza w Poznaniu, p. 165.

58. Strategia rozwoju miasta i gminy Mie˛dzyzdroje na lata 2004–2013, 2004. [Strategy of Development of Mie˛dzyzdroje 2014–2013], Rada Gminy Mie˛dzyzdroje. Strategia rozwoju wojewo´ dztwa zachodniopomorskiego do roku 2020, 2010. [Strategy of Development of the Zachodniopomorskie Province till 2020], Sejmik wojewo´ dztwa zachodniopomorskiego, Szczecin.

59. Studium uwarunkowan´ i kierunko´w zagospodarowania przestrzennego gminy Mie˛dzyzdroje, 2010a. [Study for the Spatial Plan of the Mie˛dzyzdroje commune, 2010] Mie˛dzyzdroje.

60. Studium uwarunkowan´ i kierunko´w zagospodarowania przestrzennego gminy Wolin, 2010b. [Study for the Spatial Plan

of the Wolin commune, 2010] Regionalne Biuro Gospodarki Przestrzennej Wojewo´ dztwa Zachodniopomorskiego, Szczecin.

61. The Corsini Encyclopedia of Psychology, 2010. in: Werner, I.B., Craighead, W.E., (Ed.), John Wiley and Sons, vol. 2. pp. 801–803.

62. Tomczyk, B., 2004. Turystyka na wyspie Wolin- stan i perspektywy. in: Wyspa Wolin. Zro´wnowa_zony rozwo´ j w strefie przygranicznej, Oficyna In Plus, Szczecin, pp. 145–159.

63. To_zsamos´c´ regionalna mieszkan´ co´w wojewo´ dztwa zachodniopomorskiego, 2011. [Regional identity of inhabitants of Western Pomerania Province], Urza¸d Marszałkowski, Szczecin.

64. Trosiak, C., 2008. O to_zsamos´ci regionalnej za Ziemiach Zachodnich I Po´ łnocnych', Siedlisko no5. in: Mazur, Z., Tujdowski, M., (Eds.), Instytut Zachodni, Poznan´ , pp. 1–7.

65. Tuan, Yi-fu, 1974. Topophilia: a study of environmental perception, attitudes and values, Englewood Cliffs, Prentice-Hall, New York, pp. 260.

66. Urry, J., 1994. Consuming Places. Routledge, London, pp. 257.
Wang, N., 1999. Rethinking authenticity in tourism experience. Annals of Tourism Research. 26 (2), 349–370.

67. Warf, B., 2006. Geography and Identity. in: Warf, B., (Ed.), Encyclopedia of Geography, Sage Publications, Thousand Oaks-London-New Delhi, pp. 239–241.

68. Wolin´ ski Park Narodowy jako cze¸s´c´ dziedzictwa narodowego- ankiety, 2010. [Wolin National Park as a component of national heritage-questionnaires], Wolin National Park, Mie¸dzyzdroje.

69. Wysiedlenia, wype¸dzenia i ucieczki 1939–1945, 2008. Atlas ziem Polski. in: Sienkiewicz, W., Hryciuk, G., (Eds.), Wydawnictwo Demart, Warszawa, 240.

70. Wyspiarz niebieski ('The Blue Islander'), (2006–2011), 248–533, S´ winoujs´cie

Chapter 4

The Evaluation of Beach Recreational Index for Coastal Tourism Zone of: Delegan, Kenjeran, and Wisata Bahari Lamongan ☆

Mahmud Mustain*, Haryo Dwito Armono, Danu Tri Kurniawan

Department of Ocean Engineering, Institut Teknologi Sepuluh Nopember (ITS), Sukolilo Surabaya 60111, Indonesia

ABSTRACT

The aim of this paper is to show the comparation of recreationa index from three different coastal tourism places. The evaluation of beach recreational index for coastal tourism zone, as called Integrated Beach Value Index (IBVI), have been designed include; *beach indicators or Beach Index* (BI), *beach user's perception or Knowledge Index* (KI) and the indicator of *beach economic value or Monetery Index* (MI). These indexes have been applied for three different places that has similarity in fisiography but litle bit difference in social culture and social economical background. These places are; 1.Kenjeran Beach, Surabaya; 2.Delegan Beach, Gresik; and 3.Wisata Bahari Lamongan (WBL) Beach, Lamongan. The result of IBVI give a conclusion of the best one for WBL. This gets the average of value as the highest. The next is Kenjeran Beach, both value of BI and KI are high but the value of MI is low. On the other hand, Delegan Beach has the medium value for BI and KI while the MI is low. This is due to not yet maximum implementation in spesific management's development, still in progres of revitalisation and restructurisation. The general key factors of the index evaluation in this research are availabel of structure and infastructure. These results may be used to make analysis from the comparation amoung them, and also to make evaluation for next development.

Keywords: evaluation index, coastal tourism, Kenjeran, Delegan, WBL

INTRODUCTION

Indonesia is the biggest of archipelago country. It has more than 17.000 islands on more than 3.500 mil length. It has more than 95.181 km length of the shoreline. This is the fourth longest of shoreline country in the world. Indonesia has a rich of coastal area with the resources both biologics and non-biologics[5]. The number of the council is 492, either Kota (city council) or Kabupaten. The parts of these are on the coastal vicinity that have the otorized to make regulation and to role the coastal management decision[2,3]. The population of all coastal area in this country is around 100 million, that about 40 % of the total population. The fact is that Indonesia has a large potential of natural resources in all sectors. Specially, in the sector of coastal tourism in the long of the shoreline is as the nation resources[5]. This is the chance for us to make good coastal zone management.

This fact makes us to create the special planning for coastal zone due to the rapidly change of the coastal resource that comsequently coused by development processing and globalization effect[4,5,10]. Around 70% of the world population is life in the range of 60 km from the shoreline.

This research, especially in this paper will make analysis of evaluation index to coastal recreation area in East- Java. It includes; 1.Pantai Kenjeran, Surabaya; 2.Pantai Delegan, Gresik; and 3.Wisata Bahari Lamongan (WBL), Lamongan. Fig. 1 shows the location of each object.

Figure. 1 The location of the study

Basic Theory of Indexes in Coastal Management

Generally, identification of coastal potential resources for the recommendation of management planning is based on the simple and complex of the coastal characteristic[5,6]. Chaverri (1989) made a classification for coastal of Costa Rica using the subjective evaluation technical based on more than 100 difference characteristics[1]. Short (1993) made a classification of coastal of Australia using the risk indicator of technical surfing such as; current, tide, wind, wave, reef[12]. William (1992) constructed the classification using 50 biophysical and socioeconomically characteristics[13]. Morgan and William (1995) made identification of the coastal users, then to make the priority scale in the time of choosing the characteristic[9]. Leatherman (1997) designed the range of the classification of coastal recreation area using 50 criteria based on the physical and biological environment effect[6].

The evaluation of coastal recreation area can be used to International Certification need, also as the basis of the specification for the coastal recreation area. These relate to the water quality, social and environment information such as education programs that depend on the environment regulation. This certification shoud be regulated yearly as criteria fulfillment warranty. The certification have been regulated for technical operation, not based on the fluctuation of expectation from the users such as interesting of diving, swimming, etc.

Moreover, certification for management planning also has been applied to a coastal zone in urban area due to viability of many structures and infrastructures. Recently, special certification of management planning also starts being applied for rural area with conventional objects.

WHO and United States for Environmental Protection Agency (EPA) produced the management planning based on the integration of microbiology measurement and population investigation. Micallef and Williams (2004) developed the calculation and classification for swimming area called Bathing Area Classification System (BARE). This system uses 45 difference indicators for coastal rural and urban area[7,8]. These indicators are had been divided to several categories i.e.; safety, water quality, infrastructure, garbage that have been produced.

Pereira (2003) chosen coastal characteristic for scaling based on polling from the users[11]. Both of the coastal description and the users perceptions are had been implemented to design management planning[11]. Moreover, Micallef and Williams (2004) recommended the consideration of economical characteristic for evaluation and classification of coastal area[8]. This is to find the main factor of the quality of integration system in coastal management planning with the different condition of social economic.

a. Beach Index (BI)

This is to know how the beach is appropriate for the tourism abject. There are 36 indicators for ecological aspect in biophysical condition. These include; grain, texture, water temperature, and soon. These also include environment issue such as clean and worst. Tourism beach vicinity with the white grain is more interested than black grain. Fine texture of beach and clean from the garbage is the best tourism object. While the physical parameter of environment is temperate such as that the air temperature appropriates to the temperature of the human body, no whale animal and soon. The consideration of 38 social economical indicators is to know the available of infrastructures and services such as; parking area, restaurant, rescue team etc. Each indicator from the total of 74 evaluates the valuable of coastal area in order to make the tourism beach condition using the water and grain.

The formulation of all these recreational indexes are based on the same formulation. It is equal to the Result Index over the Maximum Index for each of BI, KI, and MI. There are differences in the number of statements among these tree categories; BI has 12, KI has 9, and MI has 16 number of statements. Therefore in this research the formulation of BI is:

BI = Result Index/ Maximum Index

Maximum Index = (i x 3) x amout of statements

$$\text{Result Index} = \sum_{i=1}^{50}\left[\sum_{j=1}^{12} P_j x3 + \sum_{j=1}^{12} P_j x2 + \sum_{j=1}^{12} P_j x1\right]_i$$

b. Knowledge Index (KI)

This is collected from the respondents as users. The questions are had been referred from Pereira S.D. in Brazil, North Irland, and Portugal[11], and from Morgan and Williams in England[9]. The questions are had been focused in three aspects:

Visitor profiles; education level, gender, affiliation etc.

Visitor motivation; reason to visit, hobby, time selection etc.

The opinion; spesific sub-object, infrastructure, services etc.

The attitude and the opinion of visitors are representative to the visitors perception called the *knowledge index* (KI). The survey have been conducted in the weekend period during the vacation term. It was the pick time of the largest of visitor number. The similar answers of the quisener have been found event using many variations and repetations of survey. The formulation of KI in this research is:

KI = Result Index/ Maximum Index

Maximum Index = (i x 3) x amout of statements

$$\text{Result Index} = \sum_{i=1}^{50}\left[\sum_{j=1}^{9} P_j x3 + \sum_{j=1}^{9} P_j x2 + \sum_{j=1}^{9} P_j x1\right]_i$$

where : i = number of respondens (50)

p = statement in the quisener

j = amount of statements (9)

The analysis of the result are shoud be done based on the frequency of the responce and the degree or level of object. The frequency analysis will give us the opinion of visitors and their attitutes, while the level of object shows;

the interested object, (2) ordinary, and (3) un-interested object. Refering to the previous research, ideally the beach recreation object is a coastal vicinity with white sand, fine grain size, shallow water, temperate of water temperature, no animal whale, clean or no garbage, not dangerous situation and condition, have appropriate infrastructure, and good basic services such as toilet, resciu team, shoping area etc.

2.3 netary Index (MI)

Economically, the indicators of interested area from the visitors are based on the some parameters. There include:

Taxt of properties

The rate of the property of land for real estate

The charge or rate of the local hotel.

A space with or whitout the coastal view and the distance from the coastal area affect the economical value to dicide this index. Each factor will be used to convert to be made the mark of the index. An example of the comparation case and numerical index value are will be standarised by using Nijkamp and Rietveld standard as the non-parametric statistic that have been defined as three classifications; high, medium, and low.

The formulation of MI in this research is:

MI = Result Index/ Maximum Index

Maximum Index = (i x 3) x amout of statements

$$\text{Result Index} = \sum_{i=1}^{50}\left[\sum_{j=1}^{16} P_j x3 + \sum_{j=1}^{16} P_j x2 + \sum_{j=1}^{16} P_j x1\right]_i$$

where : i = number of respondens (50)

p = statement in the quisener

j = amount of statements (16)

Finally, all indexes of BI, KI, and MI will have been constructed to became the matrix for the coastal valuation index as the matrix of the evaluation of beach recreational index for coastal recreation zone (IBVI). This is theoritically all of squence comceptions.

3. METHODOLOGY

a. The rsearch has several stages below;
b. Problem identification, after topic decision we identify the potensial problem, the alternative solusion, and the positive effect of the result to the users. Therefore, to deside the potensial object for this research.
c. Problem formulation, after the abject was had been desided then to chose the spesific problem for making decision for problem formulation.
d. Literature review, this is conducted to describe averithing related to this topic based on the available references. It includes; Integrated Coastal Management, Coastal Turism Area Management based on the Economical Purposes, and so on.
e. Data collection, it is started from visitation to the location, collecting data of the govermnet documentation (scandery data) such as a report of development or assestment from the related local government offices.
f. Polling of respondens informations (primary data), the special quisioner have been made for the selectec respondens.
g. Processing data, in order to make calculation for a valuation with Beach Recreational Index method.
h. Analysis the result then to make conclusion and recomendation.

4. THE RESULT

4.1 Study Area

Study area, there are three different coastal locations. These have similar characteristics in the physiography such as a morphology, but they have the difference social culture as a local potensions and economical factors. These locations are in the North area of East Java Propince; Kenjeran (Surabaya), Delegan (Gresik), and Wisata Bahari Lamongan (Lamongan).

Kenjeran-Surabaya, the object is located in East of Surabaya around 15 km from the city centre (Tugu Pahlawan). The area is abaut 10 Ha; the North is fisheries county, the East is Madura strain, the South is aqua culture, and the West is the real astate. The population is mostly Javanis and Maduranis. This objest is very easy to be attached by either public transportation or private vehicle.

Delegan-Gresik, the object is located 43 km in the North-West of city centre of Gresik. It is also can be attached by either public transportation or private vehicle at around an hour traveling from city centre. The view along of the way is very beautiful due to the combination between coastal vicinity and some interested landscape with special flora. Wisata Bahari Lamongan (WBL)-Lamongan, the object is located at Jl. Raya Dandels at the county of Paciran, around 30 km in the North-West of city centre of Lamongan. This object is just around 5 years operation. It is on area of 17 Ha. There are two difference objects near the location; Pantai Tanjung Kodok (Beach area) and Gua Istana Maharani (Hill area).

4.2 The Profiles Respondens

Each location was had been collected from 50 visitors as respondens data by quitioner. The data includes; the facilities, the services, and the perseption of respondents. There are trhee types of respondens that shown in Figs 2, 3 and 4. These related to the age, aducation level, and job scuenly.

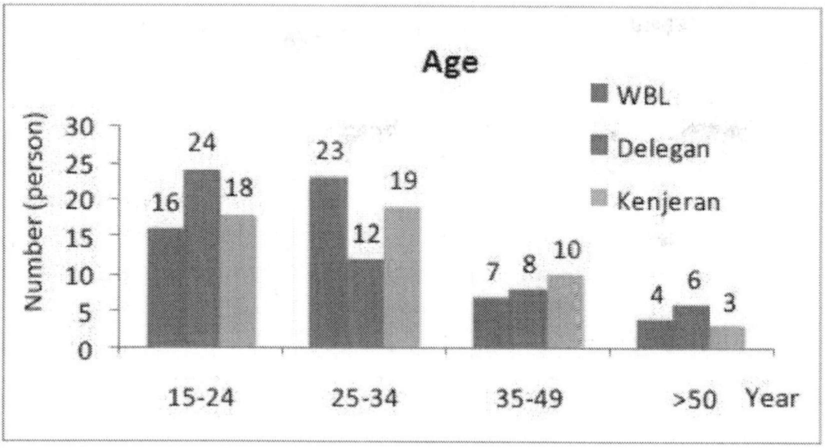

Figure 2 The age profile of respondens

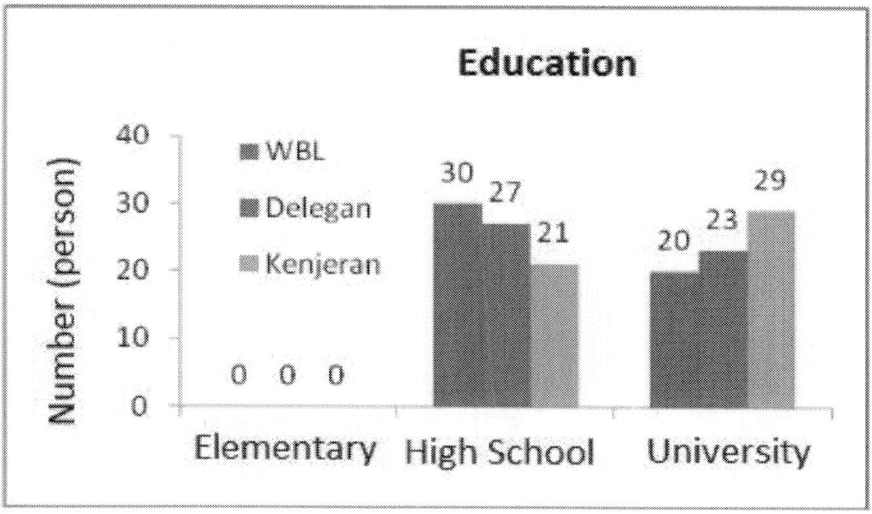

Figure 3 The education profile of respondens

4.3 The Analysis of Quitioners Result

The analysis of Beach Recreational Index uses three criteria to deside the index value. It includes; Beach Index (BI), Knowledge Index (KI), and Monetary Index (MI). BI uses the 16 categories. These are not only to describe the avaibility of the object to become the coastal recreation object but also to make analysis them. These include the consideration and evaluation of infrastructure and public sevices. For examles; parking area, restorations, toiletries, etc. The result is given in Table 1. If we refer to the range of the high criteria, BI of WBL is in the maximum range while Kenjeran is in around the minimum range of high criteria. This means that BI of WBL is much bigger than BI of Kenjeran event in the same range of high. For Delegan, the BI is in the medium range at a near maximum boundary i.e. 0.75 with range of criteria of 0.18 to 0.88. It means that the BI is not bad and not so difficult to be incresed.

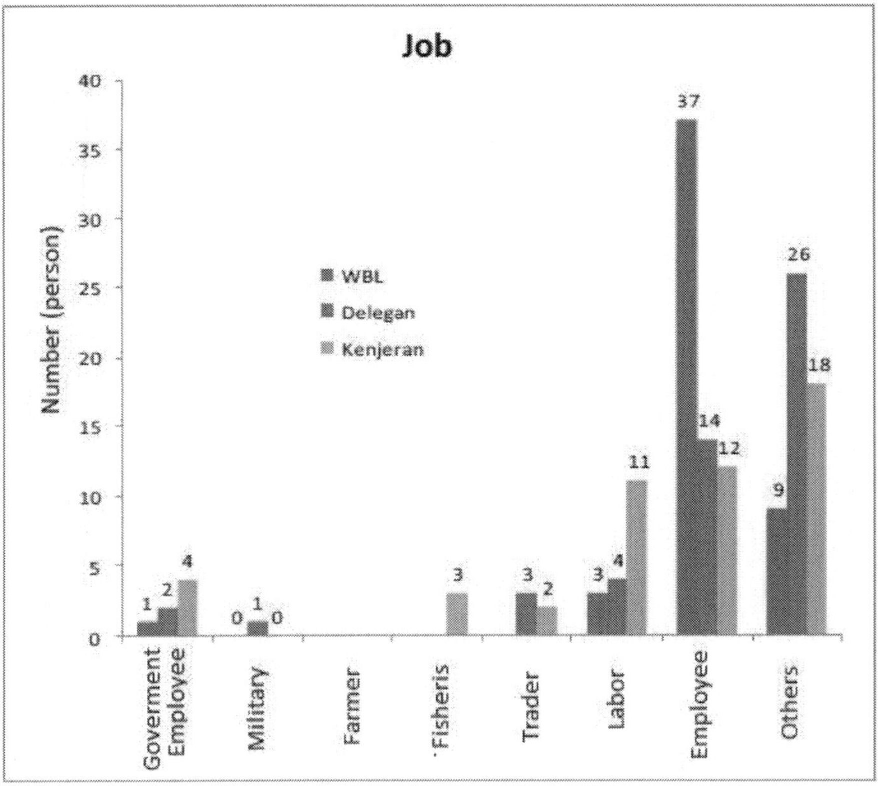

Figure 4 The job profile of respondens

KI describes the knowledge and the opinion of the respondens about the recreation object in the term of services and the avaibility information from the media such as; televition, news paper, and so on. There are 9 categories for these purposes. The result is drown in the Table 2. The Opinion of Delegan is the lowest number (zero) in the range of the low result (0.0 to 0.5). This is the serious condition for the future prespective while the WBL and Kenjeran have the savety conditions. The attitute of the respondens of all locations are in the average. The perseption of visitors gives result of medium range for Delegan and high range for both WBL and Kenjeran.

MI describes the economical effect due to the existence of the recreational objects. For example how the object could give a new income for the local comunity from the location. There are 16 categories for geting these purposes. Table 3 shows the result of the MI. In general, WBL gets the high critera except the room rates. On the othe hand, Delegan gives low criteria except room rates and property taxt. Kenjeran has the medium except property taxt and room rate.

Therefore, in the general result of Integrated Beach Value Index (IBVI) shows in Table 4. WBL shows the best condition, all indexes on the high criteria. In contrast, Delegan ilustrates the lowest condition i.e. in the medium criteria and trend to low. The most indexes of Kenjeran give a high except the MI, this shows the medium criteria.

Table 1 Beach Index (BI)

Beach	Kenjeran	Delegan	WBL
Beach Index	High 0.9	Medium 0.75	High 1

(0-0.18 = low, 0.19-0.88 = medium, 0.89-1 = high)

Table 2 Knowledge Index (KI)

Knowledge Index (KI)	Kenjeran	Delegan	WBL
Opinion	Medium 0.8	Low 0	High 1
Attitute	Medium 0.7	Medium 0.6	Medium 0.8
Perseption	High	Medium	High

(0-0.5 = low, 0.6-0.8 = medium, 0.9-1 = high)

Table 3 Monetery Index (MI)

Beach	Kenjeran	Delegan	WBL
Property tax	High 1	Medium 0.22	High 0.76
Real estate	Medium 0.54	Low 0	High 1
Room rates	Low 0	High 0.5	Low 0.16
Rank (MI)	Medium 1.54	Low 0.72	High 1.92

(0.6-0.72 = low, 0.73-1.54 = medium, 1.55-1.96 = high)

Table 4 Inegrated Beach Value Index (IBVI)

Index	Kenjeran	Delegan	WBL
Beach index (BI)	High	Medium	High
Knowledge index (KI)	High	Medium	High
Mnetary index (MI)	Medium	Low	High
Integrated beach value index (IBVI)	High	Medium	High

4.4 Discussions

Most questions of all locations figure that the most high Beach Index is for WBL. This is due to the physical atributes such as more interested of coastal zonation than the others. The avaibility of appropriate infrastrusture and good services are also give support. On the othe hand, Delegan gets BI of Medium. This is caused by the minimum of infrastrusture and unrepresentative of parking area.

In the case of KI, Wisata Bahari Lamongan (WBL) gets the best of value. This is caused by the respondens perseption about both public services and promotion media. These are very good, in the same word these are have been properly organized. In contrast, Delegan is the object that needs a serious implementation of organisation management otherwhise it still not so interested object. Eventhough Kenjeran not so good index as WBL, this object is better than Delegan. The key factor of Kenjeran is not enough promotion, not only make promotion for incidential or certain events but also need continue and appropriate inovation.

The existence of the coastal recreation objects indeed increase the social economical life especially for local community. WBL made a local community shine to have some more alternative new potential job. For examples; a new market at around WBL creates innovation of variation souvenir, some special food as product of local culture, an soon. It was also have been done by local community on around Delegan and Kenjeran event not so many in WBL. In this case WBL is the best to create the new potential job and it was have been taken by local community.

5. CONCLUSION

The results of Inegrated Beach Value Index (IBVI) are; Kenjeran: High, Delegan: Medium, and WBL: High. The dominant factors in this valuation are; Structure, infrastructure, and public services respectively. This study gives recommendation to the management of Delegan, especially to make appropriate area for parking. For the feature, this study overs to the next researcher to continue this research especially to make critical analysis to the result of questioners.

REFERENCES:

1. Chaverri, R., 1989, Coastal Mangement: The Costa Rica Experience, In: Coastal Zone'87, Vol. V. America Society of Civil En gineering, New York, Pp 5273-5285.
2. Deliyanto, B., (2001). Studi Evaluasi Dampak Pembangunan Wisata Bahari Terhadap Lansekap Lahan Pantai, Pusat Studi Indonesia Lembaga Penelitian Universitas Terbuka.

3. Isnul Aprinda, Mahmud Mustain, Arif Suroso, 2008, "Penataan Kawasan Pesisir Berdasarkan Fungsi dan Ruang Kabupaten Sambas", 2008, proceeding, Seminar Nasional Pascasarjana VIII ITS, Surabaya.
4. Kay, R., dan Alder J., (1999). Coastal Planning and Management, E & FN Spon, London.
5. La Ode Ahyar Thamrin Mufti, Mahmud Mustain, Arief Suroso, 2009, " Penilaian Ekonomi Terhadap Kerusakan Hutan Mangrove Di P ulau Kaledupa, Kepulauan Wakatobi Yang Mendukung Aspek Pengelolaan Berbasis Masyarakat Secara Terpadu", Proceding Seminar Nasional Kelautan V, Surabaya.
6. Leatherman, S.P., 1997, Beach Rating: A Methodologycal Approach. Journal of Coastal Research, 13(1), 253-258. Fort Lauderdale (Florida), ISSN 0749-0208
7. Micallef, A., (2003). Designing a Bathing Area Management Plan, Journal of Coastal Research.
8. Micallef, A., Williams, A.T., 2004, Application of a Novel Approach to Beach Classificationin the Maltese Islands. Ocean & Coastal Managmenet 47, 225-241.
9. Morgan, R. And Williams, A.T. (1995), Socio-demographyc parameter and user Priorities at Gower Beaches, UL, in Healy, M.G. and Doody, J.P., (eds), Direction in European Caostal Management, pp83-90, Cardigan, EUCC and Samara Pubishing.
10. Naomi M.D, Mahmud Mustain, Dwi Endah Kusrini, 2009, *"Studi Partisipasi Stakeholder dalam Penguatan Status Kawasan Cagar Alam*
11. *Pulau Enu dan Pulau Karang Kabupaten Kepulauan Aru",* Proceding Seminar Nasional Pascasarjana IX ITS, Surabaya.
12. .Pereira, D.S.. Landscap e Perception and Coastal Management: A Methodology to Encourage Public Perception. Journal of Coastal Research SI 39, ICS 2004 (Proceeding), ISSN 0749-0208, Brazil. 930-934.

13. Short, A.D., 1993, Beaches of The New Shouth Wales Coast. Australian Beach Safety and Mangement Project, Sydney, 298 pp. 13.Williams, A.T., (1992). Scenery and Bathing Area Registration and Evaluation of Selected Beaches Along The Coastal Province of Nador,Morocco.

Chapter 5

Alternative Strategies for Scaling Up Marine Coastal Biodiversity Conservation in Chile

Stefan Gelcich* Leornardo Peralta, C Josh Donlan, Natalio Godoy, Veronica Ortiz, Sebastian Tapia-Lewin, Camila Vargas, Andres Kein, Juan Carlos Castilla, Miriam Fernandez and Francisco Godoy

[1]*Centro de Conservación Marina, epartamento de Ecologia,Pontificia Universidad Católica de Chile, Alameda 340, Santiago 8331150, Chile*

[2]*Center of Applied Ecology and Sustainability (CAPES), Facultad de Ciencias Biológicas, Pontificia Universidad Católica de Chile, Alameda 340, Santiago 8331150, Chile*

ABSTRACT

The continued degradation of marine ecosystems, along with the ecosystem services they provide, suggest that new, innovative approaches are needed to scale up marine biodiversity protection and promote sustainable fishery practices. We synthesize information from Chile on the key processes involved in the development of alternative strategies for scaling up marine biodiversity conservation and discuss the complementarities with marine protected areas. Defined as "ancillary" marine conservation initiatives under the Convention of Biological Diversity, we suggest that these alternative strategies have the potential to capitalize on local stakeholders' participation and contribute to solving livelihood and governance issues while playing a significant role in scaling up marine conservation. We specifically focus on two recent ancillary initiatives being piloted in Chile. The development of business model innovations which could enable biodiversity benefits from territorial user rights fisheries policies and the creation of municipal conservation areas. We identify how these initiatives could eventually help scale up marine conservation, discuss opportunities and challenges from these pilot experiences and conclude with the need for developing policy frameworks and cross-scale governance approaches which formally acknowledge marine ancillary conservation measures as part of an integrated way to manage marine biodiversity. Exploring and supporting alternative complementary marine conservation strategies is particularly relevant in Chile and Latin America, if biodiversity conservation initiatives are to scale in coverage, contribute to livelihood improvement of local communities, replenish fisheries and play key roles in adaptation to climate change.

Keywords: Marine conservation Territorial user rights Navidad Artisanal fishers Municipalities Marine protected area

INTRODUCTION

In 1992 the Convention on Biological Diversity set ambitious marine conservation targets, aiming to protect at least 10% of all marine ecological regions by the year 2012 (Wells et al. 2007). This target was not achieved and re-emphasised in 2010, with a new strategic plan to enhance international efforts at stopping degradation of the world's biological heritage termed the 'Aichi Targets' (CBD 2013). The target includes developing a network of well managed Marine Protected Areas (MPAs) defined by the World Conservation Union (IUCN), as any area of inter- or sub-tidal terrain with its overlying water, associated biodiversity, historical and cultural features, which has been reserved by law or other effective means, to protect part or all of the enclosed environment (Kelleher 1999; Wells et al. 2007).

While the initial scientific focus for MPAs was on site selection (Turpie et al. 2000) and biological responses of species under protection (Micheli et al. 2004; Lester and Halpern, 2008), MPAs are now created not only to conserve seascapes and provide habitat for endangered sea-life, but also to contribute to the livelihood of local communities, to support national economies through tourism revenues, to replenish fisheries and to play key roles in adaptation to climate change, among many other functions (Watson et al. 2014). In this sense, multidisciplinary research efforts on MPAs have shifted towards defining conditions for effective management and enforcement. This has led to the recognition of the widespread occurrence of "paper-MPAs": marine protection occurs only in theory due to a lack of enforcement and management, due to the exclusive use of top-down implementation mechanisms, and lack of appropriate funding (Mora et al. 2006; Gravestock et al. 2008; Reid-Grant and Bhat 2009).

As a consequence, if marine conservation is going to effectively contribute to fulfill its multiple roles and also scale up and meet Aichi Targets, there is a pressing need to enhance the complementarities between MPAs and other management tools, by means of instruments that the Convention of Biological Diversity have termed "ancillary marine conservation initiatives" (CBD 2004). Ideally these ancillary instruments should allow participation of civil society in planning, implementing, and day-to-day management. Here, we synthesize information on two alternative strategies which could potentially serve as ancillary marine conservation instruments in Chile. We specifically explore business model innovations which could enable biodiversity benefits from territorial user rights fisheries policies (TURFs) and municipal conservation areas as ways in which Chile can scale up coastal marine conservation while securing coastal livelihoods and good governance.

REVIEW

Chilean marine coastal conservation

In Chile, the main legal tools that exist for the implementation of MPAs take the form of Natural Sanctuaries, National Monuments, Marine Parks, Marine Reserves and Multiple use MPAs (Castilla 1996; Fernández and Castilla 2005; Castilla 2008). The goal of establishing Natural Sanctuaries, Natural Monuments and Marine Parks is to preserve natural ecosystems, while also allowing for educational and research activities. Marine Parks are non-take areas in coastal or open ocean waters where marine resources are off limits to any extractive uses. In contrast, Marine Reserves allow for the rational and sustainable exploitation of resources (Fernández and Castilla 2005). Multiple Use MPAs have been the latest addition to the marine conservation policy instruments in Chile. Multiple use MPAs were first implemented with funding by the Global Environmental Facility, and are meant to act as an "umbrella-like" tool which considers the management of multiple ecosystem services within MPA boundaries (Gelcich et al. 2013).

Despite the existence of a suite of conservation tools, biodiversity conservation in Chile is underfunded, including MPAs (Castilla 2008; Gelcich et al. 2013; Waldron et al. 2013). Chile is one of four countries that are in bottom quartile of relative biodiversity conservation funding and the top quartile of threatened biodiversity (Waldron et al. 2013). The inability to direct funds to regulate and enforce MPAs has been identified as one of the main causes that lead to their failure (Mora et al. 2006; Gravestock et al. 2008). For MPAs in Chile, revenues from tourism are not sufficient to finance running costs and enforcement. For example, *Lafken Mapu Lahual*, one of the largest multiple use MPAs in continental Chile, could only achieve around 10% of running costs, in the most favorable conditions, under current management scenarios (Gelcich et al. 2013).

Despite funding constraints, the number of MPAs has increased significantly in the past years. There are currently 22 coastal areas which are protected in continental Chile (Ministerio de Medio Ambiente, 2014; Figure 1). These account for approximately 900 km^2, which considers both coastal terrestrial portions as well as marine areas within the protected sites. Yet, there are still important gaps in surface and habitat representation of protected marine ecosystems (Tognelli et al. 2009). In response, the Environmental Ministry is considering the establishment of new multiple use MPAs as a way to scale up marine conservation efforts in coastal zones. In addition to the funding constraints, this strategy will be strife with additional challenges. For example, even at the current level of MPA representation, artisanal fishers have the perception that MPAs could cause conflicts due to issues around resource access (Gelcich et al. 2009a).

Budget constraints and low stakeholder buy-in represent two premier challenges for marine biodiversity protection in Chile and elsewhere. As mentioned, we discuss two approaches, namely, the establishment of business model innovations to incentivize biodiversity benefits form TURFs and the establishment of Municipal Conservation Areas, which attempt to overcome these challenges. The driving force behind these approaches is that if properly designed and supported, these ancillary measures could complement the Chilean network of marine coastal biodiversity protected areas (Castilla 2000).

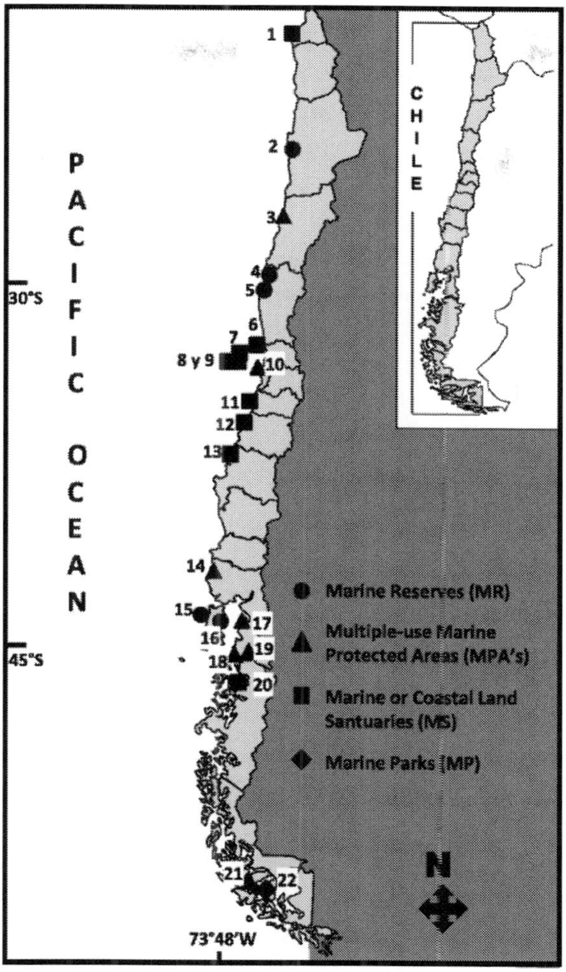

Figure 1. Map of Chile showing coastal protection areas which include marine and terrestrial ecosystems and their legal instruments. In the figure marine reserves (MR), Multiple use Marine Protected Areas (MPAs), Marine or coastal land Sanctuaries (MS) and Marine Parks (MP) are highlighted. Numbers in the figure represent: (1) MS Humedal de la Desembocadura del Río Lluta; (2) MR Bahía Moreno – La Rinconada; (3) MPA's Isla Grande Atacama; (4) MR Isla de Chañaral; (5) MR Isla Choros-Damas; (6) MS Isla Cachagua; (7) MS Roca Oceánica; (8) MS Islote Pájaro-Niño; (9) MS Peñón de Peñablanca; (10) MPA's Las Cruces; (11) MS Bosque de Calabacillo de Navidad (Municipal); (12) MS Roca de Constitución; (13) MS Lobería de Cobquecura; (14) MPA's Lafken Mapu Lahual; (15) MR Pullinque; (16) MR Putemún; (17) MPA's Fiordo Comau-San Ignacio de Huinay; (18) MPA's Pitipalena-Añihue; (19) MPA's Tic-Toc; (20) MS Estero de Quitralco; (21) MPA's Francisco Coloane; (22) MP Francisco Coloane.

Territorial user rights as possible ancillary conservation instruments

Chile is among the 10 most important countries in terms of fishery landings (FAO 2010). Thus, biodiversity conservation has to be scaled up in the context of an extremely productive marine coastal ecosystem (Thiel et al. 2007). In the last 5 years the total aggregated industrial and artisanal wild species landings, have ranged, around 4.5 million tons/year (Castilla, 2010). Approximately 50% of marine landings are based on artisanal fisheries operating in coastal zones (approximately <10 miles offshore; Castilla 2010). In Chile, to be classified as an "artisanal fisher", vessels must not exceed 18 m in length and have a maximum of up to 50 gross register tons (Fisheries and Aquaculture Law N° 18 892; Castilla 2010). Within coastal zones, the artisanal fleet supplies a significant fraction of high-valued finfish, small-pelagic fish, benthic invertebrate, and algal resources. For instance, between 2005 and 2012 around 32,000 artisanal fishers registered as divers extracted about 245,000 tons year^{-1} of benthic resources (excluding algae), worth approximately 340 $US million year^{-1}. (SERNAPESCA Servicio Nacional de Pesca 2005). Artisanal fin-fishers, during 2005–2012, extracted an average of 1.14 million tons year^{-1}of high value fin-fish and small-pelagic species, worth around 2465 US$/ton (SERNAPESCA Servicio Nacional de Pesca 2005). The artisanal fleet also lands about 373000 tons year^{-1} of algae worth 250$US million year^{-1} (SERNAPESCA Servicio Nacional de Pesca 2005). The legal framework (Fisheries and Aquaculture Law N° 18.892) that regulates fisheries in Chile provides a series of management policies such as marine zoning, regulating mobility of fleets, allocating exclusive territorial users rights for fisheries (TURFs), establishing management plans and establishing catch share systems for fully exploited species (Castilla 2010).

The TURFs policy implemented in Chile takes the form of Management and Exploitation Area for Benthic Resource policy (Castilla 1994). Through this policy the undersecretary of fisheries assigns exclusive access diving rights to fisher organizations (Castilla et al. 1998; Gelcich et al. 2010). The rational for establishing these user rights is based on common-property and co-management theories, which establish that securing access and sharing control over resources can create incentives for sustainable institutional arrangements among fishers, who will then manage and fish collectively and sustainably (Ostrom and Schlager 1996). In addition, it should contribute to more effective enforcement by increasing their likelihood of compliance (Jentoft et al. 1998). In order to be granted a TURF fisher organizations must actively engage in developing (with the help of biological consultants) official management plans. They are also responsible for surveillance and enforcement of anti-poaching measures (Castilla 2008). The first TURF was informally established in 1989 (Castilla et al. 1998). Currently in Chile there are 707 TURFs legally allocated to fisher organizations (Sernapesca 2010). Although there is heterogeneity in their performance, they account for more than 1,100 km^2 of the nearshore seascape, with an average size of approximately 100 h and an average distance between them of 4–10 km (Gelcich et al. 2010). TURFs are created and assessed considering economically important benthic species such as the carnivorous muricid gastropod *Concholepas concholepas* (considered in 80% of TURFs), key-hole limpets, *Fissurella spp*. (70%), and the red sea urchin *Loxechinus albus* (30%) and more recently for algae species (Castilla et al. 1998; Castilla et al. 2007). Biological studies of TURFs policy have proclaimed substantial increases of abundances and sizes of managed species within well enforced TURFs in comparison with open-access areas (Castilla et al. 1998).

In Chile the discussions on the role of TURFs as ancillary conservation instruments have historically been absent of the debate on scaling up marine conservation (Gelcich et al. 2011). Recently, the biodiversity conservation implications of TURFs began to be assessed scientifically (Gelcich et al. 2008, 2012). Results of these studies showed that TURFs can sustain densities and biomasses comparable to that of a fully protected no-take MPA for target species. TURFs also had significantly higher reef fish and macroinvertebrate species richness, biomass, and density compared with open-access areas (Figure 2). Furthermore, results suggest that the level of enforcement, aimed at preventing poaching in TURFs, is associated with biological diversity (Figure 2). Despite these benefits provided by TURFs for the subtidal communities, No-take MPA of around 20 hectares in size show higher density, biomass, and species richness of macroinvertebrates and reef fishes than TURFs, which indicate TURFs cannot replace no-take MPAs (Gelcich et al. 2012, Figure 2). In essence, biodiversity benefits from well managed TURFs are somewhere in between open access and No-take MPA levels.

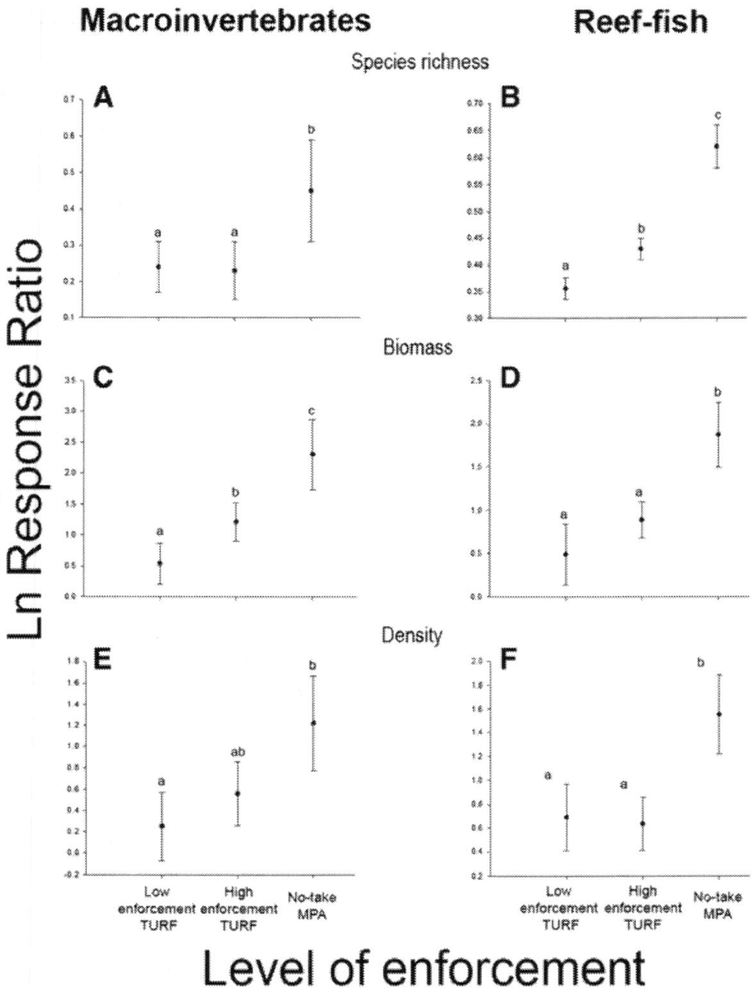

Figure 2. Response ratios for: (A) macroinvertebrate (≥3 cm) species richness, (C) macroinvertebrate biomass, (E) macroinvertebrate density, (B) reef-fish species richness, (D) reef-fish biomass, and (F) reef-fish density across different levels of enforcement of TURFs and the no-take MPA of Las Cruces. Response ratios represent the ratio of the specific measured variable in a given level of enforcement relative to the measured average value of that variable across 7 neighboring open-access areas. Hence, positive values of response ratios indicate greater species richness, biomass or density relative to open-access sites. 95% confidence intervals of response ratios were calculated with a 10,000-iteration bias-corrected and accelerated bootstrapping procedure. Letters above bars in each section indicate differences. The figure was constructed by the authors using the data reported in Gelcich et al. 2012.

The presence of TURFs, although probably providing fewer conservation outcomes than a no-take MPA, does allow marine conservation to scale-up in size. In continental Chile there is potential for conservation in 1,100 km2 of TURFs, with enforcement and surveillance in hands of artisanal fishers. There is also empirical evidence that the participation of fishers in TURFs has increased their stewardship capacity (Gelcich et al. 2008). Thus, it becomes imperative to assess and recognize the value of these ancillary measures in terms of effective marine conservation and to establish financial mechanisms to optimize their contribution (Gelcich et al. 2011).

An interesting model that is currently under consideration and analysis which could improve the potential of TURFs to scale up biodiversity conservation relates to the establishment of no-take zones within a portion of a TURF (Gelcich and Donlan, in press). Initial pilot case studies are being established to test this alternative, with promising preliminary results (Gelcich and Donlan, in press; www.advancedconservation.org). The aimed outcome of generating no-take MPAs within TURFs is a scalable program that can provide a supplementary revenue stream to fishers in exchange for management actions that produce enforced and verified biodiversity benefits, while also promoting sustainable fisheries (Gelcich and Donlan, in press).

Depending on how biodiversity incentive schemes are established, optimizing biodiversity within TURFs through establishing no-take zones within their boundaries can become a cost effective measure. TURFs hold the potential to enable at least two business model innovations that could improve fishers' livelihoods with biodiversity benefits. First, outcomes from conservation practices, such as the implementation of no-take areas within TURFs, could be commoditized and sold as credits in offset type markets (Gelcich et al. 2011, Donlan 2015). Because there are on-going marine and coastal impacts from the private and public sector throughout Chile, we anticipate opportunities for marine biodiversity offset programs. In fact, the Chilean government has recently modified important aspects of environmental impact assessment policy to allow offsetting. In addition, it is in the process of revising the offset framework within a new biodiversity and protected area policy. Second, biodiversity benefits from TURFs could be integrated and add value to products within emerging sustainable seafood markets (e.g., traceable seafood products with biodiversity benefits). Both these alternatives build upon basic conditions enabled by the TURFs, however to develop these concepts further requires the construction of learning platforms, collaborative demonstration-scale experimental trials where fishers, managers and scientists can co-construct the necessary knowledge.

Informing on innovative conservation approaches which are based on TURFs and their potential to scale up marine biodiversity conservation will require an understanding of biodiversity response, the development of financing strategies, which must be tailored to local realities, an understanding of the demand for biodiversity credits and fishers behavioral responses. New interdisciplinary approaches will be critical to solve these emerging research frontiers.

Municipal conservation sanctuaries as possible ancillary conservation instruments

Since the 1992 UN Conference on Environment and Development of Rio, international and national approaches to conservation have had to harmonize with social needs and the development agenda (Adams 2001). As a part of this process the concept of good governance of protected areas has evolved and is currently associated with fair methods, negotiation processes and the search for consensus among a plurality of actors and interested parties. This should result in building better bases for societal decision making (Borrini-Feyerabend et al. 2004). Concomitantly research and international *fora* have identified challenges related to promoting broad participation of interested parties in the management of protected areas and the fair and equitable distribution of the benefits derived from conservation (Adams 2001). A basic element of these transitions is the principle of subsidiarity which holds that a larger and greater body should not exercise functions which can be carried out efficiently by one smaller, but rather the former should support the latter and help to coordinate its activity with the activities of the whole community. As a way to promote the abovementioned principles in conservation, the IV World Congress of Nature emitted resolution 4.037 regarding municipal conservation areas. This resolution urges the director general of IUCN to "recognize the value of Municipal Conservation Areas for biodiversity conservation and their importance for increasing the effectiveness of protected area systems; and develop the capacity to promote and assist initiatives involving members in the creation and strengthening of Municipal Conservation Areas through technical and financial support"

(UICN Unión Internacional para la Conservación de la Naturaleza 2009 44–46). While municipal conservation areas can become a burden for municipalities who have no access to support networks and financial resources, the resolution establishes municipal conservation areas can become a suitable instrument for attempting to reach the goals of biodiversity conservation by linking conservation efforts with local needs. They also have the potential to contribute to efforts to generate more polycentric types of management for protected area systems (Ostrom et al. 1961).

In Latin America municipal conservation areas are relatively new phenomena, with only a few experiences which focus primarily on terrestrial environments from which to draw lessons (GTZ cooperación técnica alemana 2010). In Chile there is great potential to develop municipal conservation areas; there are 345 municipalities out of which 101 are in coastal zones (Gelcich et al. 2009b). In addition, Chile already counts with a National Municipality legislation named the The Ley Orgánica de Municipalidades de Chile, which requires that every municipality in the country establish a four year planning and management instrument, including comprehensive regulations, addressing the sustainability, rational management and conservation of resources and the environment within the municipality territories and/or ecosystems (Gelcich et al. 2009b).

Unfortunately, despite municipalities' potential in scaling up marine conservation, so far there is no clear legal approach through which municipalities can obtain management rights and funding to conserve marine areas, or is there a formal recognition of the role they can play for scaling up marine biodiversity conservation. Thus there is an urgent need to develop and support learning platforms and strengthen management skills, to highlight the real potential of marine municipal conservation areas.

One example of municipal marine conservation in Chile has been developed in Navidad, Libertador Bernardo O´Higgins Region, Chile. In Navidad, the local municipality, fisher unions and university academics began the process of applying for a no-take marine reserve in a united way in 2005. The initiative relies on collaboration of this diverse set of stakeholders that operate at different levels. The process of selecting sites and developing the base-line studies took approximately 4 years and was a collaborative process between fishers, academics and municipal officials. In addition, fisher union presidents, academics and the local council held meetings with all artisanal fishers with the objective of modifying and validating the results from site selection and base-lines (Oyanedel et al. 2015). Outreach of the final decisions towards the broader civil society was led by the municipality and fisher unions (Figure 3). In addition educational activities surrounding the sanctuary initiative have been a constant element (Figure 3).

Figure 3. Photo panel showing A) Informative panels that explain to the community and tourist about the marine conservation initiative carried by local fishers, the municipality and scientist of the Universidad Catolica de Chile. B**)** Opening ceremony where the Marine Municipal Sanctuary area of Navidad was officially established. **C**) Educational field activity developed by the Municipality for local students to learn about marine conservation. **D**) Outset of the "Alliance of Coastal Municipalities for Sustainable Development", where eight costal municipalities committed to work together to re-vindicate their rights to manage and conserve marine coastal environments.

The municipality-fisher-university collaboration brought new ideas regarding ways to conserve marine biodiversity, but more importantly, it brought in networks of contacts that help members of the fisheries union to access non-local institutions and resources (Gelcich et al. 2010). An analysis of communication, support and information networks of the Navidad initiative show how these three aspects were enhanced through the process of application and design of the marine sanctuary (Oyanedel et al., 2015). The fact that expanding and linking networks of exchange helps facilitate integrated and inclusive coastal management is not new. Tomkins et al. (2002) in Trinidad and Tobago show how those networks spread across national and international boundaries in ways that would have been hard for locals to do on their own. An important factor in developing this successfully is the presence of "linking organizations" between local actors and other scales of organizations. Linking organizations provide opportunities by bringing in resources, knowledge and other incentives to engage in marine conservation networks (Folke et al. 2005).

The Navidad marine Sanctuary is currently the first marine municipal conservation area in Chile. It was formally approved in 2013, eight years after the initial idea emerged from the fishers, municipality officials and university scientists (Oyanedel et al. 2015). The emergence and consolidation of the sanctuary demanded several collective action tasks (Ostrom, 1990). These include communication, coordination of actions, mechanisms for solving conflicts and information sharing. The organizational structure and the development of trust and reciprocity between stakeholders which emerged during the 8 years were instrumental in confronting this challenge (Oyanedel et al. 2015). In essence, these bottom-up partnerships for marine conservation did not emerge automatically in response to potential benefits. There was a need to address transaction costs associated to participation, communication and engagement disparities at local scales (Oyanedel et al. 2015).

The Navidad marine Sanctuary is not an isolated experience; there is an increasing interest for marine municipal conservation areas from mayors, local councilors and communities in Chile. For instance in Ancud, located in the Chiloe island, the local municipality with help from NGOs and the support of fisher communities established a binding municipal rule (Ordenanza Municipal in Spanish) to regulate access and visiting frequency to a penguin colony in its coastal zone. In addition, in other areas of Chile, artisanal fisher Unions have managed to summon several stakeholders to begin to locally work on zoning coastal areas, which can then be validated by Municipalities.

Experiences such as the ones above, in which coastal municipalities have engaged in marine biodiversity conservation, have triggered interest and support of other municipalities in Chile. As a product of this, on the 27th – 28th of January 2011 eight municipality mayors with their environmental specialists got together and formed the "Alianza de Municipios Costeros por el Desarollo Sostenible" (Aliance of Coastal Municipalities for Sustainable Development: Figure 3) an alliance which manifests its right to be recognized as managers of marine biodiversity and therefore a right to participate in designing marine public policy and coastal spatial planning (Gelcich et al. 2011). The alliance is open to new coastal municipalities in Chile wishing to collaborate. Ideally, this initiative could help inform legislation to acknowledge municipal marine protected areas as part of a future national system of protected areas and engage the international conservation community to support and finance the creation of capacities and social capital for their development.

DISCUSSION AND CONCLUSION

Policy decisions regarding conservation and management of natural resources directly affect fishers and coastal inhabitants' ability to maintain their communities and livelihood. Traditionally, these policies have tended to be driven in a top-down manner and consider the biophysical aspects of management decisions first and the socioeconomic impacts secondly. Experience from Chile, presented in this paper, suggests that in order to effectively scale up marine conservation practices in order to achieve both international commitments and effective coastal governance it might be more effective to emphasize socioeconomic aspects and develop mechanisms that acknowledge participation of civil society early in the process.

Participation in MPA processes tends to promote desired changes by having community members empowered with a sense of ownership (Pollnac and Pomeroy, 2005). Research which has explored experiences from municipal conservation areas show that the collaborative work aimed at facilitating dialogue and solutions, which led to the establishment and implementation of the Navidad MPA, paved the way for a consolidation process (Oyanedel et al. 2015). However the widespread implementation of municipal conservation areas is challenging, common interests do not necessarily lead to successful implementation, funding constraints and the lack of capacities could jeopardize these initiatives. However, external agents can support these processes by facilitating dialogue and finding solutions (White and Runge, 1995).

Creating market (i.e., non-regulatory) incentives to establish enforced, no-take zones within TURFs presents an opportunity to integrate livelihood improvement and marine biodiversity protection (Gelcich and Donlan, in press). In principle, using this strategy as an ancillary conservation instrument could apply to any small-scale TURFs fishery across the globe. Yet, not all TURFs are created equally, and they have been created mainly as fisheries management tools. Thus, exactly how a program can incentivize human behavioral changes to produce biodiversity benefits and how those benefits can be financed will depend on local social-ecological conditions. Important conditions which must be considered include the existence of latent biodiversity benefits and fishers' capacity to enforce TURF areas. Local enforcement by fishers not only requires a system of local governance and financial incentives, but also cross-scale linkages with regional and national institutions that can develop and support effective enforcement (Cudney-Bueno and X. Basurto 2009; Gelcich et al. 2010).

In essence, it is important to highlight that the recognition of ancillary conservation measures for marine conservation must be coupled and integrated to existing centralized experiences to achieve success. The dichotomy between top-down and bottom-up processes must turn into a synergy between these processes (Carcamo and Gaymer, 2013). In Chile evidence shows there is a strong potential for bottom-up processes in marine conservation, but there is also a need for top-down steering and guidance (Gelcich et al. 2009a). Cross scale linkages between scales of governance are crucial for the provision of services, goods and infrastructure related to the protection and enhancement of marine biodiversity, the economic development of community enterprises, and the political representation of communities (Grilo 2011).

Marine conservation in Chile is undergoing an important transformation. The newly elected government of Michelle Bachelet which took office in March 2014 has highlighted the need to establish a "National Service of Protected Areas and Biodiversity" which will gather and concentrate the diverse conservation instruments in both land and sea, under an integrated environmental governance scheme. This represents a key opportunity to highlight the role that ancillary conservation instruments such as TURFs and Municipal conservation areas could have in scaling up and managing marine biodiversity. This overarching national framework for conservation must also begin to consider the development of innovative financing mechanisms such as voluntary biodiversity offset programs (McKenney and JM Kiesecker 2010; Donlan, 2015) and market based mechanisms to finance conservation (Lester et al. 2013). The ancillary efforts presented here have the potential to enhance the overall governance of coastal natural resources and improve livelihood security of the people involved. Developing these issues further will require a planned research agenda including multidisciplinary teams, who must inform political discussions to generate the learning process for effective conservation of coastal zones.

By synthesizing information on the key processes involved in the development of pilot ancillary conservation initiatives, our goal has been to establish a better understanding of the broader foundations for innovating on ways in which marine conservation outcomes can be scaled up while accounting for local governance and livelihood realities. Important paradigm shifts which must be considered in a marine conservation agenda for Chile and Latin America must consider an evolution from conservation run by central government to that run by multiple-partners, including local communities, the private sector and NGOs. An evolution from conservation paid for by taxpayers to that paid for from multiple sources to achieve self-sustaining (Gelcich et al. 2013). Furthermore, a major shift is needed, signaling that marine coastal conservation should not only benefit visitors and tourists, but also, local communities, which assume the opportunity costs of conservation. We advocate that acknowledging and developing the concept of ancillary conservation measures is one way of including these paradigm shifts in marine conservation policies, not only in Chile, but more broadly in Latin America.

DECLARATIONS

Acknowledgements

We thank Financiamiento Basal FB-0002, nucleo-milenio initiative RC130024 and NC-120086 from the Ministerio de Economia, Fondecyt 3150138, The David and Lucile Packard Foundation and The Pew Marine Conservation Fellowship Program.

Competing interests

The authors declare that they have no competing interests.

Authors' contributions

All authors contributed to the review, drafted, read and approved the final manuscript.

REFERENCES

1. Adams, WM. 2001. Green Development. Environment and Sustainability in the Third World. London: Routledge.
2. Borrini-Feyerabend, G, A Kothari, G Oviedo. 2004. Indigenous and Local Communities and Protected Areas: Towards Equity and Enhanced Conservation. IUCN, Gland, Switzerland and Cambridge, UK. xviii + 111pp.
3. Carcamo, FP, and CF Gaymer. 2013. Interactions between spatially explicit conservation and management measures: Implications for the Governance of marine protected areas. Environmental Management 52: 1355–1368. 10.1007/s00267-013-0167-9.View Article
4. Castilla, JC. 1994. The Chilean small-scale benthic shellfisheries and the institutionalization of new management practices. Ecology International Bulletin 21: 47–63.
5. Castilla, JC. 1996. La futura red chilena de parques y reservas marinas y los conceptos de conservación, preservación y manejo en la legislación nacional. Revista Chilena de Historia Natural 69: 253–270.
6. Castilla, JC. 2000. Roles of experimental marine ecology in coastal management and conservation. Journal of Experimental Marine Biology and Ecology 250: 3–21.View Article
7. Castilla, JC. 2008. El océano y la conservación en Chile: Los eternos olvidados. Estudios Públicos 112: 206–217.
8. Castilla, JC. 2010. Fisheries in Chile: small-pelagics, management, rights and sea zoning. Bulletin of Marine Science 86: 221–234.
9. Castilla, JC, S Gelcich, and O Defeo. 2007. Successes, lessons, and projections from experience in marine benthic invertebrate artisanal fisheries in Chile. Pages 25–42 in Fisheries

management: progress toward sustainability, ed. T. McClanahan and J.C. Castilla, Oxford: Blackwell.

10. Castilla, JC, PH Manríquez, J Alvarado, A Rosson, C Pino, C Espóz, R Soto, and O Oliva Defeo. 1998. Artisanal Caletas: as units of production and co-managers of benthic invertebrates in Chile. Canadian Journal of Fisheries and Aquatic Sciences (Special Publication) 125: 407–413.
11. CBD. 2004. CBD Technical Series number 13. Montreal, Quebec: Canada. Secretariat of the Convention on Biological Diversity: technical advice on the establishment and management of a national system of marine and coastal protected areas.
12. CBD. 2013. CBD Quick guides to the Aichi Biodiversity Targets, Retrieved 25 August, 2013, from http://www.cbd.int/sp/targets/.
13. Cudney-Bueno, R., and X. Basurto. 2009. Lack of cross-scale linkages reduces robustness of community based fisheries management. Public Library of Science ONE 4: doi:10.1371/journal.pone.0006253.
14. Donlan, CJ (ed) 2015. Proactive Strategies for Protecting Species: Pre-listing Conservation and the Endangered Species Act. Berkeley, CA: University of California Press.
15. FAO. 2010. El estado mundial de La pesca y Acuicultura. Departamento de Pesca y Acuicultura de la FAO. Roma: Organización de las Naciones Unidas para la Agricultura y la Alimentación.
16. Fernández, M, and JC Castilla. 2005. Marine conservation in Chile: historical perspective, lessons, and challenges. Conservation Biology 19: 1752–1762.View Article
17. Folke, C, T Hahn, P Olsson, and J Norberg. 2005. Adaptive governance of social–ecological systems. Annual Review of Environment and Resources 30: 441–473.View Article
18. Gelcich, S, N Godoy, L Prado, and JC Castilla. 2008. Add-on conservation benefits of marine territorial user rights fishery policies in central Chile. Ecological Applications 18: 273–81.View Article

19. Gelcich, S, N Godoy, and JC Castilla. 2009a. Artisanal fishers' perceptions regarding coastal co-management policies in Chile and their potentials to scale-up marine biodiversity conservation. Ocean and Coastal Management 52: 424–432.View Article
20. Gelcich, S, O Defeo, O Iribarne, G Carpio, R DuBois, S Horta, JP Isacch, N Godoy, PC Peñaloza, and JC Castilla. 2009b. Marine ecosystem-based management in the Southern Cone of South America: Stakeholder perceptions and lessons for implementation. Marine Policy 33: 801–806.View Article
21. Gelcich, S, TP Hughes, P Olsson, C Folke, O Defeo, M Fernández, S Foale, LH Gunderson, C Rodríguez-Sickerth, M Scheffer, RS Steneck, and JC Castilla. 2010. Navigating transformations in governance of Chilean marine coastal resources. Proceedings of the National Academy of Sciences 107: 16794–16799.View Article
22. Gelcich, S, L Peralta, C Gonzalez, A Camano, M Fernandez, and JC Castilla. 2011. Scaling-up marine coastal biodiversity conservation in Chile: A call to support and develop ancillary measures and innovative financing approaches. In Successful and Failed Experiences in Biodiversity Conservation: Lessons and policy recommendations from the American continent, ed. E Figueroa, 199–220. Santiago: Editorial Universitaria.
23. Gelcich, S, M Fernandez, N Godoy, A Canepa, L Prado, and JC Castilla. 2012. Territorial user rights for fisheries as ancillary instruments for marine coastal conservation in Chile. Conservation Biology 26: 1005–1015.View Article
24. Gelcich, S, F Amar, A Valdebenito, JC Castilla, M Fernandez, C Godoy, and D Biggs. 2013. Financing marine protected areas through visitor fees: Insights from tourists willingness to pay in Chile. AMBIO 42: 975–984.View Article
25. Gelcich S and J Donlan. In press. Incentivizing biodiversity conservation in artisanal fishing communities through territorial user rights and business model innovation. Conservation Biology.

26. Gravestock, P, MR Callum, and A Bailey. 2008. The income requirements of marine protected areas. Ocean & Coastal Management 51: 272–283.View Article

27. Grilo, C. 2011. Institutional interplay in networks of marine protected areas with community-Based Management. Coastal Management 39: 440–458.View Article

28. GTZ (cooperación técnica alemana). 2010. Áreas de conservación municipal: una oportunidad para la conservación de la biodiversidad y el desarrollo local. Reflexiones y experiencias desde América Latina. Brasilia, DF.

29. Jentoft, S, B McCay, and D Wilson. 1998. Social theory and fisheries comanagement. Marine Policy 22: 423–436.View Article

30. Kelleher, G. 1999. Guidelines for marine protected areas. IUCN, Gland, Switzerland and Cambridge UK. Available from http://data.iucn.org/dbtw-wpd/edocs/PAG-003.pdf (accessed November 2009).

31. Lester, SE, and BS Halpern. 2008. Biological responses in marine no-take reserves versus partially protected areas. Marine Ecology Progress Series 367: 49–56.View Article

32. Lester, SE, C Costello, A Rassweiler, SD Gaines, and R Deacon. 2013. Encourage sustainability by giving credit for marine protected areas in seafood certification. PLoS Biology 11(12): e1001730. 10.1371/journal.pbio.1001730.View Article

33. McKenney, BA, and JM Kiesecker. 2010. Policy Development for Biodiversity Offsets: A Review of Offset Frameworks Environmental Management 45: 165–176.

34. Micheli, F, BS Halpern, W Botsford, and RR Warner. 2004. Trajectories and correlates of community change in no-take marine reserves. Ecological Applications 14: 1709–1723.View Article

35. Ministerio de Medio Ambiente. 2014. Access at http://www.mma.gob.cl/1304/w3-channel.html.

36. Mora, C, S Andrefouet, MJ Costello, C Kranenburg, A Rollo, J Veron, KJ Gaston, and RA Myers. 2006. Coral reefs and the

global network of marine protected areas. Science 312: 1750–1751.View Article

37. Ostrom, V, CM Tiebout, and R Warren. 1961. The organization of government in metropolitan areas: a theoretical inquiry. American Political Science Review 55: 831–842.View Article
38. Ostrom, E, and E Schlager. 1996. The formation of property rights. In Rights to nature: ecological, economic, cultural and political principals of institutions for the environment, ed. S Hanna, C Folke, and K Maler, 127–157. Washington: Island Press.
39. Ostrom, E. 1990. Governing the Commons: The Evolution of Institutions for Collective Action. New York, NY: Cambridge University Press.View Article
40. Oyanedel, R, A Marín, JC Castilla, and S Gelcich. 2005. Emergence and consolidation of Marine Protected Areas: Insights from two contrasting bottom-up initiatives in central
41. Pollnac, RB, and RS Pomeroy. 2005. Factors influencing the sustainability of integrated coastal management projects in the Philippines and Indonesia. Ocean & Coastal Management 48: 233–251.View Article
42. Reid-Grant, K, and M Bhat. 2009. Financing marine protected areas in Jamaica: An exploratory study. Marine Policy 33: 128–136.View Article
43. SERNAPESCA (Servicio Nacional de Pesca). 2005–2012. Anuarios estadísticos de pesca. Servicio Nacional de Pesca. Ministerio de Economía Fomento y Reconstrucción. Available from: http://www.sernapesca.cl
44. Sernapesca. 2010. Areas de Manejo en Chile.. http://www.sernapesca.cl/. Accessed 16 Mar 2015.
45. Thiel, M, EC Macaya, E Acuña, WE Arntz, H Bastias, K Brokordt, PA Camus, JC Castilla, LR Castro, M Cortés, CP Dumont, R Escribano, M Fernandez, JA Gajardo, CF Gaymer, I Gomez, AE González, HE González, PA Haye, JE Illanes, JL Iriarte, DA Lancellotti, G Luna-Jorquera, C Luxoro, PA Manriquez, V Marín, P Muñoz, SA Navarrete, E Perez, E Poulin, J Sellanes, HH

Sepúlveda, W Stotz, F Tala, CA AThomas, JA Vasquez Vargas, and A Vega. 2007. The Humboldt Current System of Northern-Central Chile: Oceanographic Processes. Ecological Interactions and Socioeconomic Feedback. Oceanography and Marine Biology: An Annual Review 45(195–344): 45.

46. Tognelli, M, M Fernandez, and P Marquet. 2009. Assessing the performance of the existing and proposed network of marine protected areas to conserve marine biodiversity in Chile. Biological Conservation 142: 3147–3153.View Article

47. Tomkins, E, WN Adger, and K Brown. 2002. Institutional networks for inclusive coastal zone management in Trinidad and Tobago. Environment and Planning 34: 1095–111.View Article

48. Turpie, JK, LE Beckley, and SM Katua. 2000. Biogeography and the selection of priority areas for conservation of South African coastal fishes. Biological Conservation 92: 59–72.View Article

49. UICN (Unión Internacional para la Conservación de la Naturaleza). 2009. Resoluciones y Recomendaciones. Disponible en <http://data.iucn.org/dbtw-wpd/edocs/WCC-4th-005-Es. pdf>1. Consultado el 26 de julio de 2010. Gland: uicn.

50. Waldron, A, AO Mooer, DC Miller, N Nibbelinka, D Redding, TS Kuhnc, JT Roberts, and JL Gittlemana. 2013. Targeting global conservation funding to limit immediate biodiversity declines. Proceedings of the National Academy of Sciences of the United States of America 110: 12144–12148.View Article

51. Watson, JEM, N Dudley, DB Segan, and M Hockings. 2014. The performance and potential of protected areas. Nature 515: 67–73.View Article

52. Wells, S, N Burgues, and A Ngusaru. 2007. Towards the 2012 marine protected area targets in eastern Africa. Ocean & Coastal Management 50: 67–83.View Article

53. White, TA, and CF Runge. 1995. The emergence and evolution of collective action: Lessons from watershed management in Haiti. World Development 23: 1683–1698.

Chapter 6

Climate Change Impacts on Indonesian Coastal Areas

Muhammad Zikra[a,*], Suntoyo[a], Lukijanto[b]

[a] *Department of Ocean Engineering, Institut Teknologi Sepuluh Nopember (ITS), Kampus ITS, Surabaya 60111, INDONESIA*

[b] *Badan Pengkajian dan Penerapan Teknologi (BPPT), M.H. Thamrin 8, Jakarta 10340, INDONESIA*

ABSTRACT

The problems in coastal area related with climate change have continued to receive a high level of attention. Elevate average of sea level, variation in significant wave height and increased ocean temperature are linked to global climate in many ways. Thus, climate variability and future climate change should become a major interest for engineer, stakeholders, and decision makers, especially for developing strategies for mitigation and adaptation for future coastal development. The objective of this paper is to compile data and information on the potential impacts of climate change to Indonesian coastal area due to global warming from various studies. This paper focus on the impacts of climate change to sea level, wave climate and sea water temperature.

Keywords: Climate change; sea level; wave climate; sea water temperature

1. INTRODUCTION

Climate change could affect coastal and ocean environments in different of ways. Coasts are sensitive to sea level rise, changes in the frequency and intensity of wind speed, increases in significant wave height, and increases in ocean temperature. Moreover, rising concentrations of carbon dioxide (CO2) are causing the oceans to absorb more of the gas and become more acidic. This rising acidity could have significant impacts on coastal environments and marine ecosystems. The impacts of climate change are likely to worsen many problems that coastal areas already face. Coastal erosion, coastal flooding, and water pollution are common problems that affect man-made infrastructure and ecosystems in coastal areas.

As the world's largest archipelagic state, Indonesia have more than 17,500 islands with over 81,000 kilometers of coastline. The coastline of Indonesia is highly populated because around 220 million Indonesians reside within 100 km of the coast, and of these over 150 million people rely on marine resources for their livelihoods (WRI, 2001). All activities in the coastal and ocean, such as marine transportation, offshore industry, naval industry, resource extraction, fish cultivation and tourism become an important part of Indonesian economy grow. Therefore, information about the effect of climate change on coastal areas is the major knowledge for engineer, stakeholders, or decision makers to develop strategies for mitigation and adaptation for future coastal development.

1.1 Sea level

Global warming as a result of the effects of greenhouse gases, have an impact on rise in sea levels. IPCC (2007) reported that the sea level has risen by an average of 2.5 millimeters annually. As an archipelago country with over 80.000 kilometers of coastlines, Indonesia is very vulnerable to sea level rise. Recently, there are several remote sensing technologies that can monitor the condition of the oceans continuously. Satellite altimeter technology is one of technique to monitor sea level change. During the past two decades, observations from satellite altimeters have demonstrated dramatic descriptions of sea level variability with higher spatial resolution than the traditional tide gauges (Ami et al, 2012).

This section described the projected increase in sea level based on the data of satellite altimeter. This study used the altimeter data from Jason-2 over Indonesia sea region from 2009 until 2012. The process and analysis of sea level data were carried out using Basic Radar Altimetry Toolbox 3.1.0 (BRAT). Monitoring of sea level rise is done for 4 years period (2009-2012) in 4 locations, which are Medan, Pemangkat, Ambon, and Manokwari. Defian (2012) reported that the highest sea level rise is in Manokwari 14.1 mm / year, and the lowest is in Ambon at 1,175 mm / year. Table 1 shows the statistics of sea level anomaly for Jason-2 after correcting for sea level height.

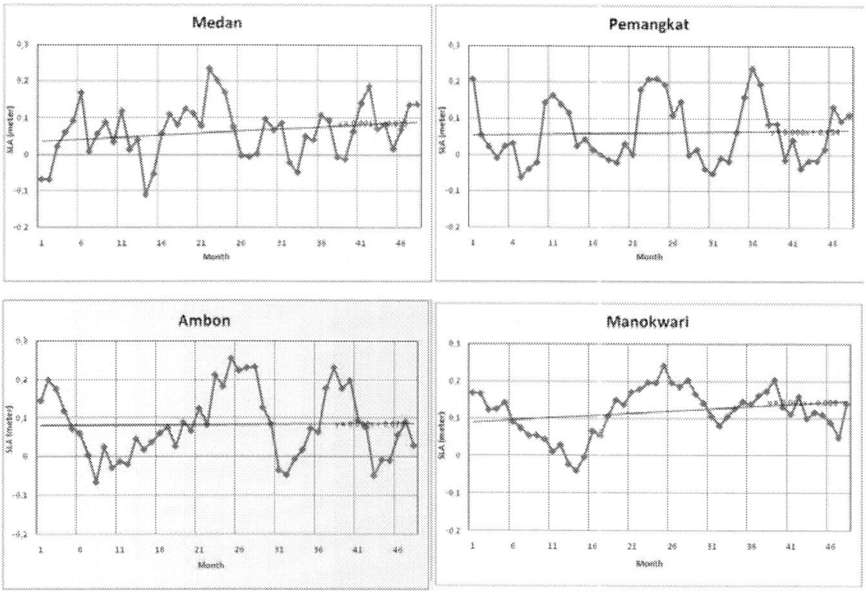

Figure 1. Trend of sea level rise

Table 1. Sea level trend based on altimeter data.

Location	Sea level anomaly (mm/year)
Medan	12.9
Pemangkat	3.53
Ambon	1.18
Manokwari	14.1

1.2 Variation in significant wave height from 1984-2003 around Indonesia sea

The global ocean wave climate has long been of interest to the ocean engineering community because of the need for accurate operational wave data for applications such as vessel design, design of offshore and coastal structures or naval operations. Recently, there has been a major interest in wave climate changes as a result of global warming. Therefore, studies on predicting the effect of global warming on ocean wave climate are required. The objectives of this section are to analyze the variability of significant wave height for the 20 year period 1984-2003.

This study describes the 20 year global significant wave height simulation derived from the Japan Meteorology Agency/Meteorology Research Institute (JMA/MRI)-AGCM3.2 wind climate data. The wind climate data were input into ocean wave model WAM with a global grid of spacing 1° in latitude by 1° in longitude. In situ wind and wave data sets from National Data Buoy Center (NDBC)-National Oceanic and Atmospheric Administration (NOAA) database were used to evaluate the model accuracy. The validation showed good agreement both wind and waves data (Zikra et al, 2015). Normalita et al (2014) reported that from 1984-2003, the trend of annual mean significant wave height was depend on observation location and time (season). The increasing trend of significant wave height is in North Natuna, Banda Aceh, North Papua and South Jogjakarta with the range 0.38 – 0.75 cm/year, respectively as shown in Table 2.

Table 2. Trend of significant wave height from 1984-200

Point	Significant wave height (cm/year)
1. 110^{0} longitude, 5^{0} latitude (North Natuna)	0.38
2. 90^{0} longitude, 5^{0} latitude (Banda Aceh)	0.52
3. 110^{0} longitude, -10^{0} latitude (South Jogyakarta)	0.75
4. 140^{0} longitude, 10^{0} latitude (North Papua)	0.75

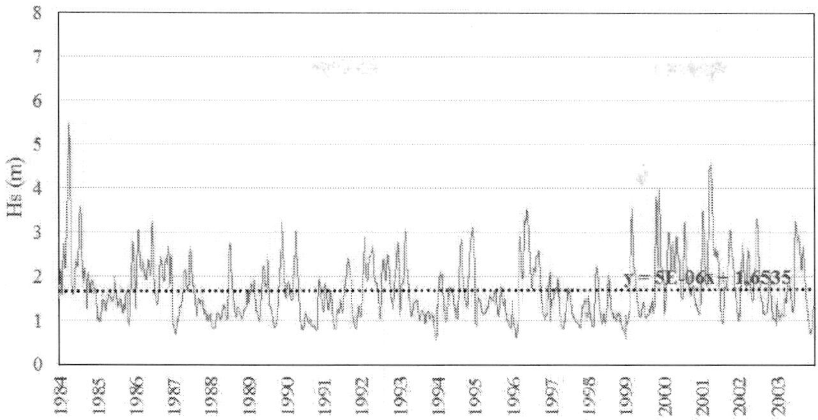

Figure 2. Temporal variation in the mean of s gnificant wave height (Location : South Cina Sea – North Natuna (Longitude 110°, Latitude 5°)

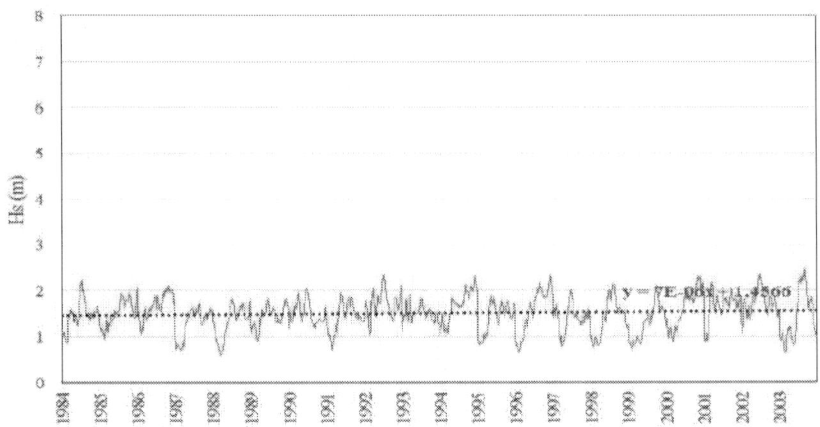

Figure 3. Temporal variation in the mean of significant wave height (Location : Indian Ocean – West Aceh (Longitude 90°, Latitude 5°)

1.3 Sea water temperature

Coastal waters have warmed during the last century, and are very likely to continue to warm by as much as 4 to 8°F in the 21st century (IPCC, 2007). This warming may lead to big changes in oceanic circulation patterns and salinity, affecting species that inhabit these areas. Along with the intensification of the process of global warming, the intensity of El Nino and La Nina is also increasing (Timmermann et al. (1999). The El Nino event in 1997/1998 caused a long dry season and caused coral bleaching in Indonesia. At the time of occurrence of La Nina in 1999, Indonesia experienced an increase in high rainfall, and high rise sea level of 20cm to 30cm, causing flooding in most Indonesian coastal territory areas (PEACE, 2007).

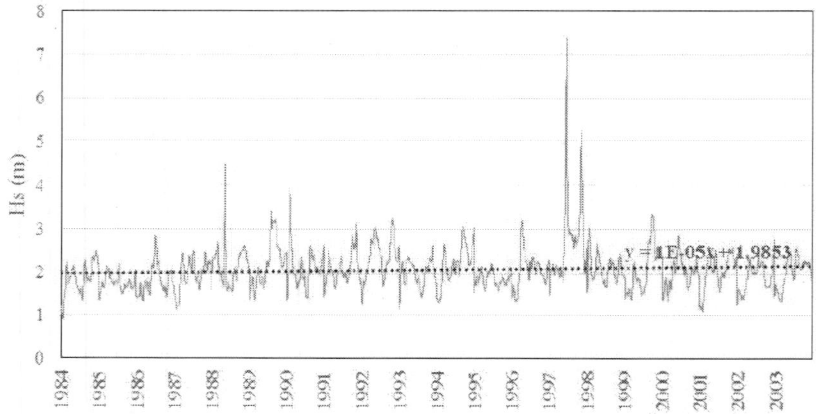

Figure 4. Temporal variation in the mean of significant wave height (Location : Pacific Ocean – North Papua (Longitude 140°, Latitude 10°)

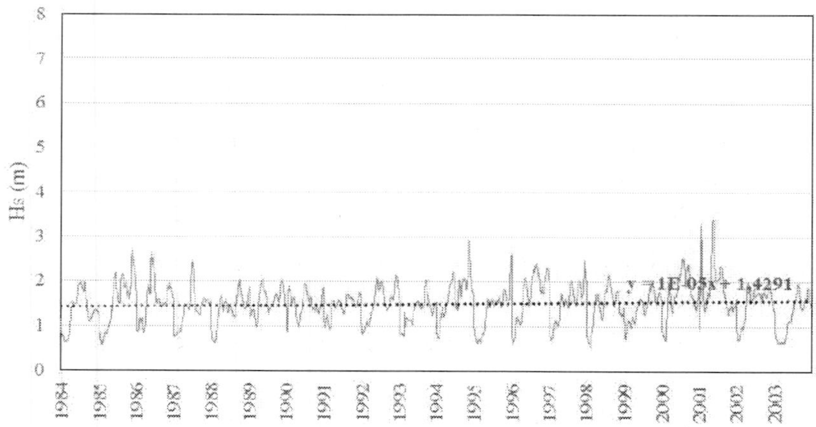

Figure 5. Temporal variation in the mean of significant wave height (Location : Indian Ocean – South Yogyakarta (Longitude 110°, Latitude -10°)

This section specifically analyze about the pattern of temperature and salinity changes upon the Pacific Ocean caused by El Nino and La Nina based on study by Firra et al. (2013). This study used temperature and salinity data from TRITON buoy located at precisely 0°N 138°E south west side of the Pacific Ocean for 5 year period (2005- 2010) as shown in Figure 6. This TRITON buoy was developed and deployed in the western equatorial Pacific and Indian Ocean since 1998 by Japan Agency for Marine-Earth Science and Technology. Basically, the purpose of this project is to observe ocean and atmosphere in the western tropical Pacific Ocean for better understanding of climate variability involving the El Nino and La Nina phenomena (JAMSTEC, 2010). The study concluded that the change in the sea surface temperature in the south west side of the Pacific Ocean towards the warm side, which explains the stronger and more frequent El Nino observed during 2005-2010 as shown in Figure 7.

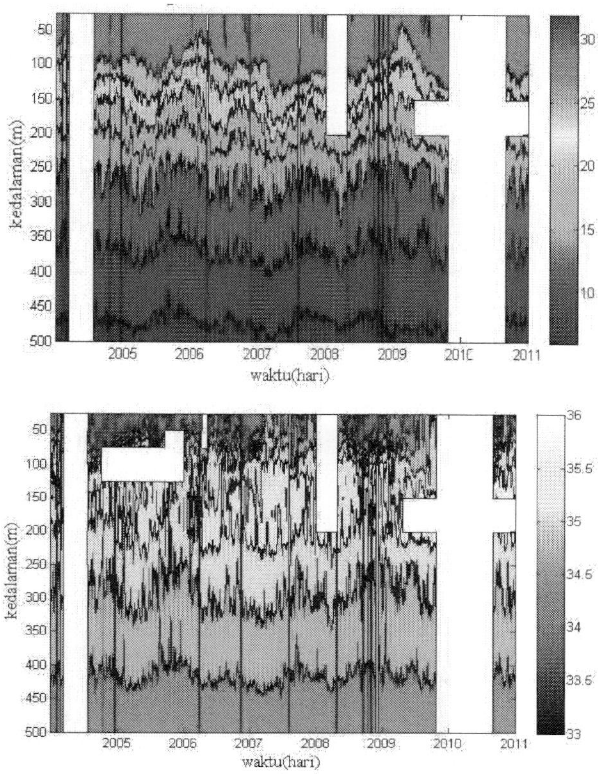

Figure 6. (top) temperature; (below) salinity.

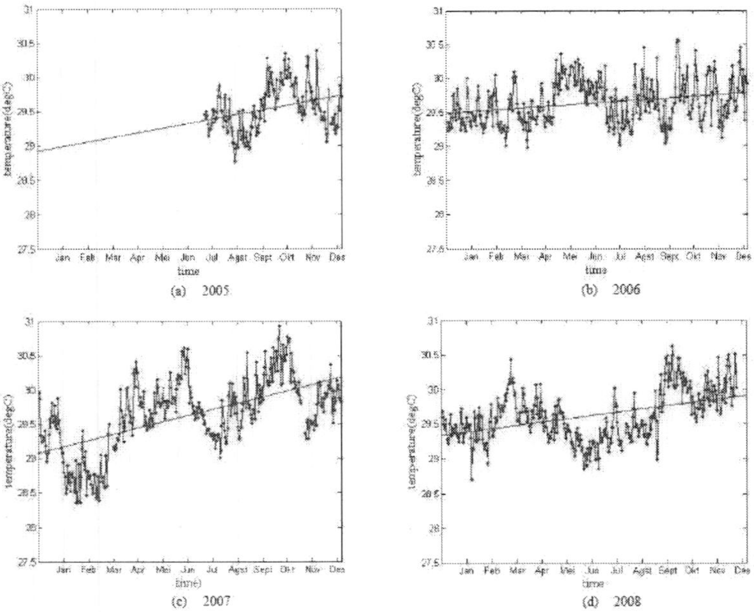

Figure 7. Trend of sea surface temperature from 2005-2010

CONCLUSIONS

As an archipelago with over 17.500 islands and over 81,000 kilometers of coastline, Indonesia coastal area is very vulnerable to climate change. Increasing trend of sea level rise, warmer ocean temperature and increased of significant wave height are among a few example of what climate change may bring to Indonesia. These problems should received serious attention from government to develop adaptation and mitigation plans for future development associated with coastal areas.

ACKNOWLEDGEMENTS

The authors would like to express their sincere thank to Japan Meteorology Agency/Meteorology Research Institute (JMA/MRI) and Japan Agency for Marine-Earth Science and Technology (JAMSTEC) for the use of data. This study is being supported by

DIKTI Research Program under Grand Number 016452.27/IT2.7/PN.01.00/2014.

REFERENCES

1. Ami Hassan Md Din, Kamaludin Mohd Omar, Marc Naeije and Sahrum Ses. Long-term sea level change in the Malaysian seas from multi- mission altimetry data. *International Journal of Physical Sciences.;* 2012; Vol. 7(10), pp. 1694 - 1712
2. Defian Naturezza. *Analisis Kenaikan Muka Air Laut Berbasis Sistem Informasi Geografis Menggunakan Satelit Altimetri, Undergraduate Report.* Institut Teknologi Sepuluh Nopember, Surabaya; 2009. p. 106
3. Fu, L.L. and Cazenave A.. *Altimetry and Earth Science, A Handbook of Techniques and Applications*. Vol. 69 of International Geophysics Series. Academic Press, London; 2001.
4. Firra Hasita, Muhammad Zikra and Suntoyo. Analisa variasi temperatur dan salinitas air laut di perairan Samudra Pasifik akibat pengaruh El- Nino dan La-Nina. *Jurnal Teknik Pomits;* 2013. Vol. 2, No. 2, pp. 2301-9271
5. IPCC. *Contribution of Working Group I to the Fourth Assessment Report of the Intergovernmental Panel on Climate Change* . Cambridge, United Kingdom and New York, NY, USA: Cambridge University Press; 2007.
6. JAMSTEC. *R/V Mirai Cruise Report.* Annual Report; 2010. p. 105
7. Muhammad Zikra, Noriaki Hashimoto, Kodama Mitsuyasu and Kriyo Sambodho. Monthly Variations of Global Wave Climate due to Global Warming. *Jurnal Teknologi*; 2015. Vol. 74. No. 5. DOI: http://dx.doi.org/10.11113/jt.v74.4637

8. Normalita Sari, Muhammad Zikra, Suntoyo. Analisa Variasi Musiman Tinggi Gelombang Laut Secara Global Selama Periode 1984 -2003,
9. *Proceeding Seminar Nasional Pascasarjana XIV- ITS.* Surabaya. Indonesia; 2014.
10. PEACE. *Indonesia and Climate Change: Current Status and Policies*; 2007. p. 90.
11. Timmermann, A., J. Oberhuber, A. Bacher, M. Esch, M. Latif, and E. Roeckner. Increased El Niño frequency in a climate model forced by future greenhouse warming. *Nature;* 1999. 398. pp. 694-697.
12. WRI. *World Resources 2000-2001: People and Ecosystems*: The Fraying Web of Life; 2001.

Chapter 7

Coastal Erosion in China under the Condition of Global Climate Change and Measures for its Prevention

Feng Cai[a,*], Xianze Su[a], Jianhui Liu[b], Bing Li[b], Gang Lei[a]

[a] *Third Institute of Oceanography State Oceanic Administration, Xiamen 361005, China*

[b] *College of Marine Geo-Science, Ocean University of China, Qingdao 266003, China*

ABSTRACT

The general characteristics of coastal erosion in China are described in terms of the regional geography, the form of erosion, the causes of erosion, and the challenges we are facing. The paper highlights the relationship between coastal erosion and sea level rises, storm waves and tides, and the influence of global climate changes on coastal erosion along the coastal zone of China. The response of the risk of coastal erosion in China to climate changes has obvious regional diversity. Research into and the forecasting of the effects of climate changes on coastal erosion are systemic work involving the natural environment, social economy, and alongshore engineering projects in the global system. Facing global warming and continual enhancement of coastal erosion, suggestions for basic theoretical study, prevention technology, management system assurance, and strengthening the legal system are presented here.

Keywords: Global climate changes; Sea level rise; Coastal erosion; China

1. INTRODUCTION

The global average sea level has been rising over the last 100 years, and with global warming the annual rate of sea level rise is expected to be two to five times the present rate. By 2100, the sea level is projected to be approximately 50 cm higher than it is today. Two-thirds of the world's major cities, which contain 60% of the population and have higher levels of economic development, are located in coastal zones [1]. For instance, more than 8 out of 10 Australians (85%) live within 50 km of the coastline [2]. Anticipated climate changes will greatly amplify risks to coastal populations, and by the end of this century, the global sea level rise will lead to the inundation of low-lying coastal regions, inducing more frequent flooding during storm surges and beach erosion [3] and [4]. Saltwater could penetrate further up rivers and estuaries and infiltrate coastal aquifers and contaminate urban water supplies.

Coastal erosion is a global problem. If the sea level rises in tandem with the occurrence of greater and more frequent storms, coastal flooding and erosion problems will become exacerbated in vulnerable coastal areas [5]. At least 70% of the sandy beaches around the world are recessional [6]. In the United States of America, approximately 86% of the United States east coast barrier beaches (excluding evolving spit areas) have experienced erosion during the past 100 years [7]. Widespread erosion is also well documented in California [8] and the Gulf of Mexico [9].

In China, coastal erosion has become a major concern for future socio-economic developments in coastal cities. Shoreline retreat in low-lying areas around the Shandong Peninsula has been greatly accelerated [10]. A maximum coastal retreat of 300 m $year^{-1}$ has been estimated at the Luanhe River mouth and an average erosion of 25 m $year^{-1}$ for its offshore sandbars accompanied by a rapid decrease in sandbar size [11].

A sea level rise and the negative environmental effects caused by human activities aggravate the risk of coastal erosion and increase the environmental burden in these areas day by day. The trend of global warming is hard to reverse, and it appears that the global climate change will not be under control in the near future. Therefore, research institutions, coastal zone management organizations and local ocean departments in the coastal states or areas have been focusing their attention on coastal erosion and disaster prevention.

The aim of this paper is to present the primary analysis of the effects of climate change on coastal erosion in China and to propose prevention measures. However, many unknowns require further research.

2. CHARACTERISTICS OF COASTAL EROSION

The continental coastline of China extends for about 18,000 km, from the Yalu River mouth in the north to the Beilun River mouth in the south, forming a southeastward convex arc. The total length of island coastline is about 14,000 km. Coastal erosion, which is widespread in China, is distributed over one-third of the coastline with a preliminary estimate suggesting that 46% of the Bohai Sea coastline, 49% of the Yellow Sea coastline, 44% of the East China Sea coastline (including the Taiwan Island coastline), and 21% of the South China Sea coastline (including the Hainan Island coastline) suffered erosion [12]. Coastal erosion along the shoreline in China is significantly affected by several factors that reflect complicated phenomena and processes. In this section, three aspects of coastal erosion are briefly discussed.

2.1. Large-scale regional variation in coastal erosion

The Meso-Cenozoic evolution formed a regional tectonic framework characterized by a landform higher in the west than in the east. China's coastline faces the world's largest marginal sea. In the coastal zone from northwest to southeast lie the Yanshan uplift belt, Xialiaohe-North China subsidence belt, Jiaoliao uplift belt, South Yellow Sea-North Jiangsu subsidence belt and Zhe Min Yue Gui uplift belt [13] (Fig. 1), and the climate varies obviously from north to south. The macro tectonic background not only determined the coastal evolution and features, but also influenced the sediment budgets and stability of the coastal zone. Thus, understanding the tectonic background would greatly help researchers studying large-scale regional variations in coastal erosion.

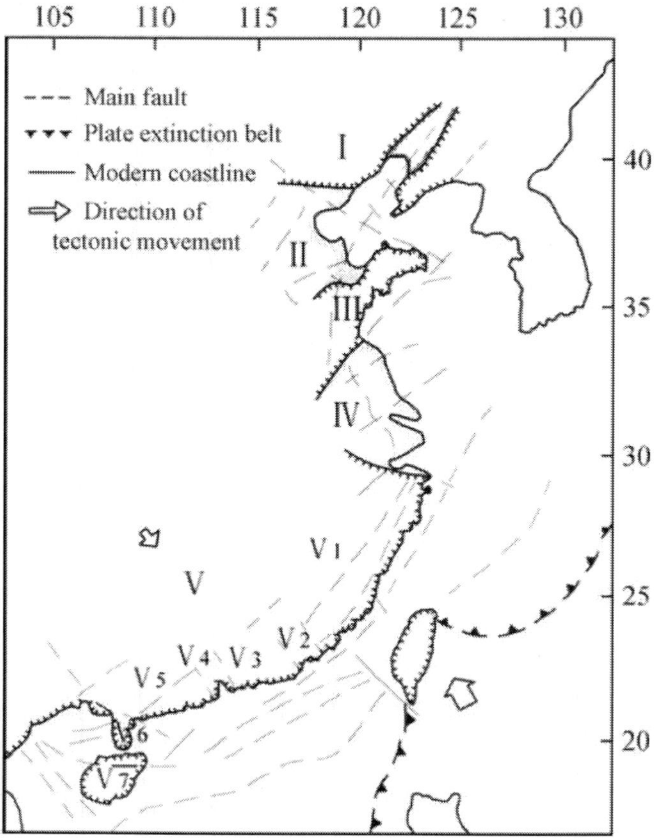

Figure 1. A sketch showing the tectonic movement of Mesozonic–Neozonic eras along the coastal zone of China. I, Yanshan uplift belt; II, Xialiaohe-North China subsidence belt; III, Jiaoliao uplift belt; IV, South Yellow sea-Northern Jiangsu subsidence belt; V, Zhe Min Yue Gui uplift belt; V_1, Zhemin uplift belt; V_2, Hanjiang delta fault depression; V_3, East Guangdong fault uplift; V_4, Pearl River delta fault depression; V_5, West Guangdong-South Guangxi fault uplift; V_6, Leiqiong fault depression; V_7, Centre-South Qiong fault uplift.

The wide distribution and various degrees of erosion exhibit different features in different tectonic belts.

1) *Subsidence belts:* the change in the river watershed, reduction of the sediment budget and change in offshore submarine geomorphology due to natural processes and human activities

result in large scale and widespread muddy coastal erosion. Erosion chiefly occurred on the abandoned sub-delta muddy coast of the modern Yellow River estuary, the abandoned Yellow River delta and Qionggang and Lusi muddy coast in North Jiangsu, and part of the muddy coast of the Yangtze River estuary.

2) *Uplift belts:* hurricanes and storm surges, sand mining and reclamation form uplift belts with the erosion of sandy coast, typically on the east coast of Liaodong Bay near Xiongyue, the coast of Qinhuangdao, the coast of Penglai and southern Rizhao in Shandong Province, the east coast of Xiamen Island, the coast of Shuidong port in Guangdong Province, and the coast of the Nandu River estuary on Hainan Island.

Different climatic zones are distributed from north to south along China's coast, having different climate impacts, biological successions and human activities. Coastal erosion for different climatic zones is roughly described as follows:

(1)*Tropical and subtropical zones:* erosion occurring along southern China and the nearby islands has destroyed the coral reefs; e.g., over 80% of coral reefs have been destroyed on Hainan Island (The State Oceanic Administration, 1996). There has also been loss and degradation of wetland and the recession and disappearance of mangroves; e.g., the area of mangrove forest has been reduced by 65% since the l950s [14].

(2)*Temperate zone:* erosion having occurred in the coastal plains, delta plains and estuarine river banks of this climatic zone can be described as shoreline retreat and downward cutting of the shore face. An investigation found that the most severe coastal erosion was along the northern coastline.

2.2. Main types of coastal erosion

Coastal erosion has different spatial and temporal forms (Fig. 2). There are three major spatial forms of coastal erosion: (1) coastline retreat, which occurs dominantly for soft coast (comprising quaternary sediment and eluvial sediments of red soil weathering crust and barrier-lagoons) without protection measures like seawall engineering, (2) landward movement of the zero meter depth contour, which is caused by beach surface incision, usually occurring on a coast with a seawall, and (3) downward erosion of the lower beach in the sub-tidal zone by tidal current with the upper flat maintaining its original shape. The former two cases are widespread, and the latter case, for instance, was found on the tidal flat of Jinshazui in the north of Hangzhou Bay.

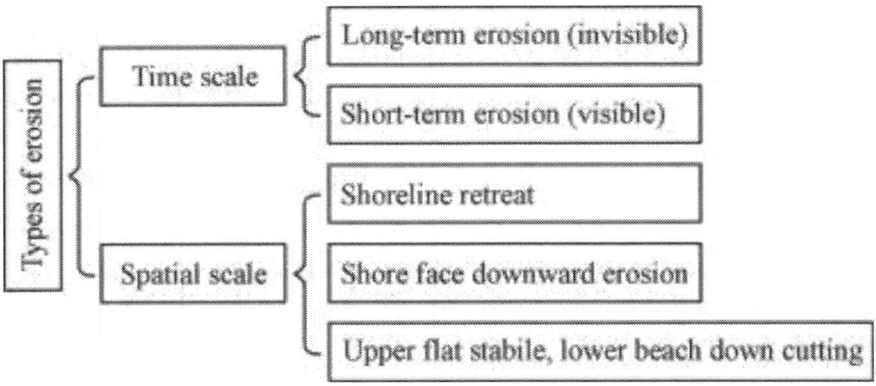

Figure 2. Forms of coastal erosion.

Erosion can also be divided into two types in terms of time scale: (1) Long-term erosion (invisible) is the permanent change in the shoreline position due to events such as a sea level rise, river diversion or decrease in sediment discharge, which reduce the original sediment budget. Under the circumstances of new coastal dynamics and sediment budget, erosion takes a long time and moves slowly. For instance, with the eastward movement of the Luanhe River mouth since the Holocene, the original coast of the Luanhe Delta was reworked by waves and tidal currents and as a result, the coastal plain sank and turned into a lagoon, a number of sea dikes off the coast were eroded and destroyed, and the accretion of coast switched to erosion and retreat [15]. (2) Short-term erosion (visible) can be caused by storm tides and storm surges without causing a permanent change in the shoreline position, but it brings enormous destruction. Erosion due to hurricanes and storm surges in the summer and autumn on the south coast of China is an example.

2.3. Causes of erosion and challenges faced

The causes of erosion include natural processes and human activities (Fig. 3). Erosion intensity and development depend to a high degree on the equilibrium state of coastal dynamics and beach stability. Generally, several controlling factors influence a certain coast simultaneously. With economic development and coastal exploitation since the 1950s, the impact of human activities has increased day by day. In consideration of the global sea level rise in recent decades, China is currently facing three major challenges due to the developing trend of coastal erosion.

(1)Threats of global warming and rising sea level to coastal plains

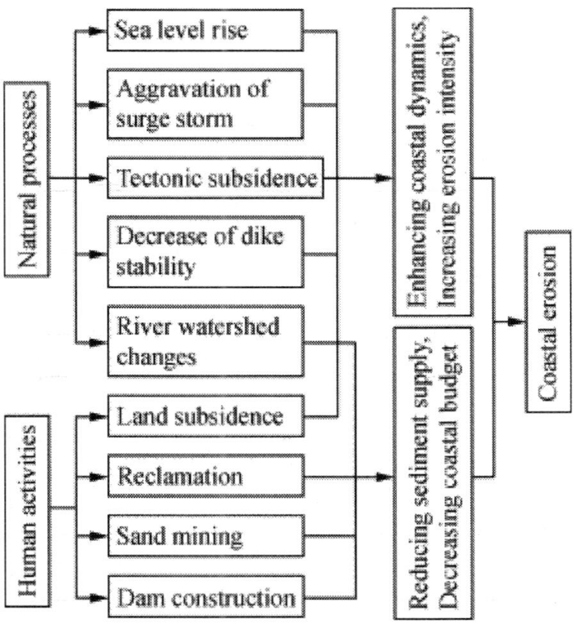

Figure 3. Sketch map of the main causes of coastal erosion.

Global warming increases the frequency and intensity of hurricanes, storm surges, and floods, while a sea level rise directly enhances ocean dynamics and causes coastline retreat. Both phenomena play important roles in coastal erosion [16]. They are mutually complementary and thus the threat is greater, especially for the coastal plains that are located in subsidence belts and fault depression basins of uplift belts, and have the most sensitive response to a sea level rise.

(2)Variation of sediment discharge

In recent decades, in the pursuit of enormous economic benefits, the main river watersheds have undergone increased exploitation, especially a series of large engineering projects such as that which returned cultivated land to forest in western China, the Three Gorges project, the Xiaolangdi project, and the south-to-north water transfer project. As a result, sediment discharge into the sea has decreased sharply. According to an updated estimate presented by Chen et al. [1], the amount of sediment discharge was almost 2000 million tons per year before the 1980s and it decreased to 1000 million tons, maybe only 500–700 million tons, by the end of the 20th Century. This resulted in notable changes in erosion or accretion evolution in estuary deltas and adjacent coasts under the new dynamic environment and sediment conditions, thus reducing the sediment budget of beaches and eventually driving severe coastal erosion.

(3)Impacts of improper coastal exploitation

With the rapid development of the economy over the past two decades, the hazard of coastal erosion has become more and more serious in China because of the dominant driving force of negative environmental impacts resulting from human activities. Sand mining has been the direct cause of sandy beach erosion in uplift belts in recent years. Over-exploitation of groundwater under a coastal plain resulted in land subsidence and triggered immediate saltwater erosion. Reclamation, reservoir and other unreasonable coastal engineering projects effectively reduced the coastal sediment discharge. All the above human activities, which have severe negative impacts on the environment and arrest the increasing attention of coastal researchers, are indirect causes of coastline retreat. In addition, river rebuilding has also aggravated erosion in abandoned estuaries.

3. IMPACT OF GLOBAL CLIMATE CHANGE ON COASTAL EROSION

China has been experiencing accelerating coastline erosion, as have other countries. Many studies have indicated that the global climate change has a complex and extensive impact on the coastal zones [17],

[18], [19], [20] and [21]. Fig. 4 roughly shows the links between the effects of global climate change and coastal erosion. In this section, the effects of a rising sea level and storm surges are emphasized.

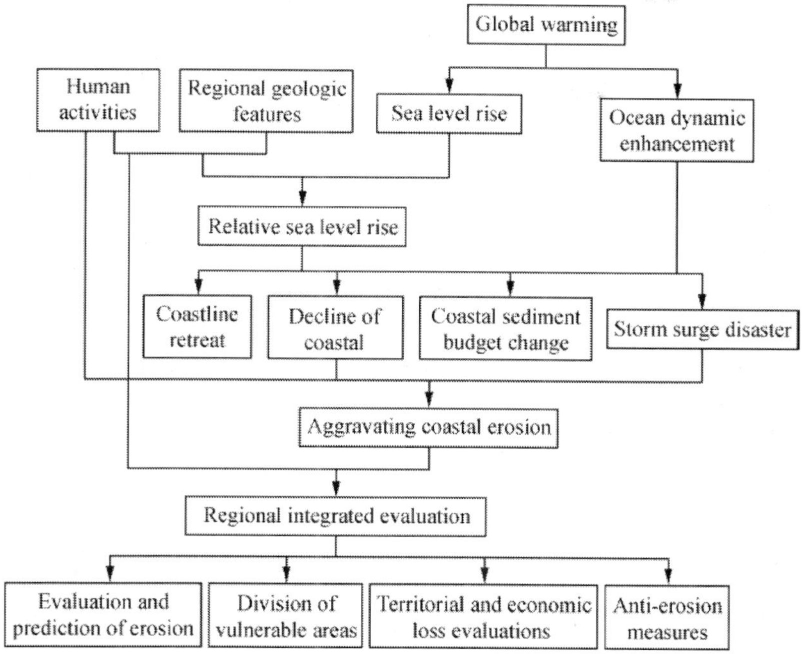

Figure 4. Effects of global changes on coastal erosion and its evaluation and prevention.

3.1. Sea level change and coastal erosion

The first model relating shoreline retreat to a rise in the regional sea level was proposed by Bruun [22]. The analysis by Bruun assumes that with a rise in sea level, the equilibrium beach profile moves upward and landward with its original shape. The upper beach is eroded owing to the landward translation of the profile, and the material eroded from the upper beach is transported immediately offshore and deposited on a nearshore bottom. The model was widely used by many researchers, and was named the Bruun Rule. Despite its simplicity and numerous assumptions, which have been criticized by some scholars [23], [24], [25], [26], [27], [28], [29], [30] and [31], the Bruun Rule has been continually developed and has worked well in many settings. The rule has wide applications: from a simple concave-downward beach profile to an underwater barrier and barrier island; from two dimensions to three dimensions, which refers to the impact of longshore sediment transport on the beach profile; from theoretical models to physical models or field demonstrations; and from sandy coast to muddy coast [32], [33], [34], [35], [36], [37], [38], [39], [40] and [41]. In addition, it also helps in designing protective measures against a sea level rise. However, the Bruun Rule cannot be applied as a common model to predict the erosion state in various coastal environments so far, and further research is needed for it to have general acceptance.

A sea level rise can be either a global sea level rise or a regional sea level rise. A global sea level rise is attributed to the thermal expansion of the upper ocean layers and melting of polar ice sheets and mountain glaciers and even to the movement of celestial bodies and sunspot activity. A regional relative sea level rise is caused not only by a global rise but also by a regional groundwater level change resulting from a monsoon, ocean current, delta accretion or engineering project, or local land subsidence, which consists of the vertical movement of the Earth's crust caused by tectogenesis and rheomorphism, sediment compaction, overexploitation of groundwater, or overloading by large buildings.

Because of the effects of local tectonic movements and human activities, the rate of the relative sea level rise varies in different areas [42]. According to a previous study [43], the rate of the regional relative sea level rise can be ten times or even a hundred times that of the global sea level rise, and as a result, research interest has switched to regional relative sea level rises since the 1990s.

There have been great achievements in the study of a sea level rise since the beginning of the new century, involving an increase in the number of monitoring stations, basic forecasting and impact assessments on the progression of sea level change, and establishment of a professional and specialized monitoring system. The China Sea Level Bulletin (Fig. 5) edited by the State Oceanic Administration makes the following remarks [44] and [45].

1) The rate of the relative sea level rise in China varied between 1.0 and 3.0 mm/year over the last 50 years, and has an accelerating trend in the last few years.
2) The rate of sea level change over the last 30 years has a general pattern of a fluctuating increase, which varies remarkably from

place to place. The rate in the south is higher than that in the north, but in the past few years, a few areas (e.g., Tianjin) in the north have also seen a continual increase.

3) The rate of sea level rise was 2.5 mm/year in 2006, and this was above the estimated mean global value of 1.8 mm/year. The value varied in different regions; the Yellow Sea had a rate of 2.5 mm/year, whereas the East China Sea had a higher rate and the Bohai Sea and South China Sea had lower rates. The China Sea Level Bulletin 2006 forecasts that the mean sea level will rise by 9–31 mm in 3–10 years.

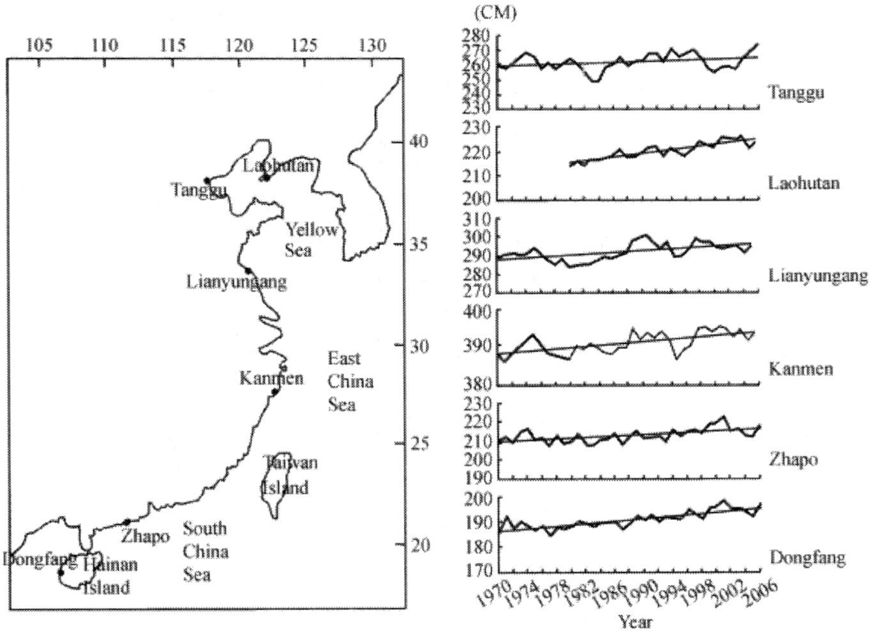

Figure 5. Sea level changes from 1970 to 2006 from observation stations along the coast of China.

From an analysis of regional differences of the main coastal plains and sea level changes in recent decades, Yang et al. [18] and [46] made a prediction of the possible sea level rise and inundated areas for the main coastal plains (Table 1).

Table 1. Prediction of the possible sea level rise and inundated area for the main coastal plains in 2050.

Region	Rate of relative sea level rise in the past few decades (mm/year)	Relative sea level rise in 2050		Expected area of inundation (hm^2)
		Rate (mm/year)	Height (cm)	
Yellow River delta	4.8	6.4–6.9	35–40	39,000
Western coastal plain of Bohai sea	2.1	11.0–11.5	65–70	21,000
North Jiangsu coastal plain	2.2	6.8–7.3	40–45	38,000–89,000
Yangtze River delta	6.6	7.9–8.4	45–50	7000–11,000
West Taiwan coastal plain*	2.0	7.0–9.5	28–38	—
Pearl River delta	2.5	4.1–4.6	25–30	2000
Hainan coastal plain	1.8	3.4–3.9	20–25	7000
Guangxi coastal plain	1.8	3.4–3.9	20–25	17,000
Hanjiang River delta	1.5	3.1–3.6	15–20	2000
Minjiang River delta	1.8	3.4–3.9	20–25	6000–15,000
Xialiaohe River delta	1.7	3.3–3.8	20–25	4000–9000

* From W.J. Chen, C.T. Kuo.

A sea level rise has been identified as the principal forcing function in large-scale shoreline retreat [47]. There are two main aspects to the response of coastal erosion to sea level change:

1) A sea level rise has direct and indirect impacts in terms of the form of erosion it promotes. Direct impacts include coastline retreat, inundation and swamping of low land on coastal plains, seawater intrusion and destruction of the seawall and drainage system. Indirect impacts are mainly caused by dynamics changes and sediment budget reductions, which induce accelerating coastal erosion. Generally, indirect impacts are more complicated.
2) Different regions have different physical geographies and levels of social–economic development, which influence the intensity and form of erosion due to a sea level rise. Therefore, different regions should be studied separately.

According to the analysis of these factors of altitude, relative sea level change (ground subsidence), coastal erosion, storm surges and coastal protection work, Ren plotted out the eight regions that were most vulnerable to coastal hazards. These regions comprised the ancient Yellow River delta, modern Yellow River delta, coastal plain in Laizhou Bay, abandoned Yellow River delta, coastal plain in North Jiangsu, Yangtse River delta, coastal plain in western Taiwan, and the Pearl River delta. It is estimated that the disaster-affected area is about 35,000 km^2.

According to the previous studies [48], [49], [50], [51], [52], [53], [54], [55], [56], [57], [58], [59], [60] and [61], we present the characteristics of erosion for the main coastal plains in Table 2. It can be concluded that a sea level rise and the environmental effects it induces are the major factors contributing to the coastal erosion.

Table 2. Characteristic erosion on the main coastal plains.

Region	Yangtze River delta	Yellow River delta	Pearl River delta	Taiwan coastal plain
Causes	Reduction of sediment supply; coastal dynamic change; mainstream swinging; flood tidal rush; tectonic subsidence; sea level rise	Reduction of sediment supply; adjustment of sedimentary dynamics; river diversion; sea level rise	Land subsidence; storm surge aggravating; weakening in defense capability of dikes; sea level rise	Reduction of sediment supply from rivers; construction of improper engineering structures; land subsidence; sand mining; sea level rise
Performance	Channel siltation, shoreline retreat, and lowland inundation	Shoreline retreat; downward erosion of beach; submerging of drilling platform; fracture of oil pipeline; destruction of seawall	Destruction of mudflat, marsh, mangrove, coral reef; devastation of seawall	Beach erosion, lowland inundation, salinization of freshwater
Examples	Over 390 km of the coast are eroded; average rate of shoreline retreat in the recent three decades in Hangzhou Bay is up to 30 cm/a	The maximum shoreline retreat reached 11 km between 1976 and 2000, with an annual loss of 420 m; the largest downward erosion depth was 10.5 m, with an annual erosion exceeding 40 cm	Several meters of shoreline retreat occurred in Shekou District between 1966 and 1979; Since 2000, the largest retreat in bridgehead of north Tangjia is 70–80 m	The coastline in Taipei retreated about 500 m between 1980 and 2003
Characteristics	Shoreline erosion is mainly located in the south margin of delta and the north part of Hangzhou Bay	With regional and unbalanced characteristics, the overall erosion trend is difficult to stop	86.7% of this delta is protected by dikes, other parts of it are mainly eroded	Relative sea level rise in the west is higher than elsewhere and make obvious coastal erosion, while some areas remain stable

3.2. Storm surge and coastal erosion

Global warming can cause a catastrophic rise in the sea level, and a warming climate is expected to raise the intensity and frequency of storm surges (tides). *The China Marine Disaster Bulletin* edited by the State Oceanic Administration reported that from 1995 to 2006, the numbers of yearly storm surges were 10, 6, 4, 7, 5, 8, 6, 8, 10, 10, 11, and 9, respectively, giving an average occurrence of 7.8 times per year. It can be seen that the frequency has increased in recent years. The relationship between climate change and the frequency of storm surges is complicated. A numerical simulation [62], case analysis [16] and the inter-annual variation in atmospheric circulation [63] and [64] show that a tropical cyclone intensity has a remarkable positive correlation with the sea surface temperature, and global climate changes have obvious effects on tropical cyclone activity.

A storm surge is a regional paroxysmal hazard that brings about the heaviest loss of life and property; for instance, storm surges caused a direct economic loss of 3.298 billion yuan (RMB) and 137 deaths in 2005 in China [65]. A surge also strongly changes the seafloor terrain, marine sediments and coastal geography; in particular, it violently swashes the shore and upper tidal flat. A strong hurricane usually induces more significant erosion than that occurring in a normal dynamic condition, and the prominent consequences from large events can remain for many years. Such destruction was reported for typhoon No. 9914 on the east coast of Xiamen Island and for typhoon No. 0307 (Imbudo) on the west coast of Guangdong Province [66] and [67].

From a study of hazards caused by storm surges in recent years, it has been noted that hurricane frequency and intensities have risen gradually. *The China Sea Level Bulletin 2006* has stated that from 2004 to 2006, more frequent and higher intensity storm surges have been responsible for the greatest property losses in the coastal areas of Zhejiang and Fujian provinces, and the losses were more serious than those in the historical records. The enormous losses were certainly related to the rapid socio-economic development of the coastal areas, but the global climate change and a sea level rise play more important roles.

Beach erosion derived from storm surges in many coastal states and areas has attracted great attention. A field survey of damage by storm surges on the coastlines and beaches in China has been undertaken in recent years, and the preliminary effects of storms on beaches in China have also been studied [66] and [68]. In future, more attention should be paid to the characteristics of beach erosion and accretion during a storm, mechanism of coastal erosion, post-disaster construction, and in particular, tendency forecasting for climate change and coastal socio-economic development.

There are several aspects to an evaluation of the impacts of climate change on coastal erosion, including the natural environment, socio-economic factors, and coastal engineering projects, which make up an integrated system. Sea level changes also play an important role in the system; however, they have temporal and spatial variations and an accurate evaluation of their contribution to coastal erosion requires further widespread ocean monitoring, data collection, and intensive study. It is concluded that different areas have great regional variations in the natural environment, ocean condition and coastal stability [69], human activities also have enormous impacts on the coastal zones, and the adaptive capacity of economically developed regions allows a more flexible response to climate change. In summary, the impacts of climate change on coastal erosion have a distinct regional variability.

4. COASTAL PROTECTION MEASURES

Global warming has accelerated the continuous development of coastal erosion, and raised the challenges in coastal zone management, exploitation and research in China, and the government and scientists have paid increasing attention to this problem. A basic theoretical study of coastal processes has been running for several decades, and has had a great deal of success in improving the integrated coastal zone management and monitoring. On the basis of recent studies, a summary of the main protection measures is presented in Fig. 6. Both long-term planning and updated protection measures are required to eliminate the adverse impacts of the hazard and guarantee sustained and harmonious development of the coastal economy.

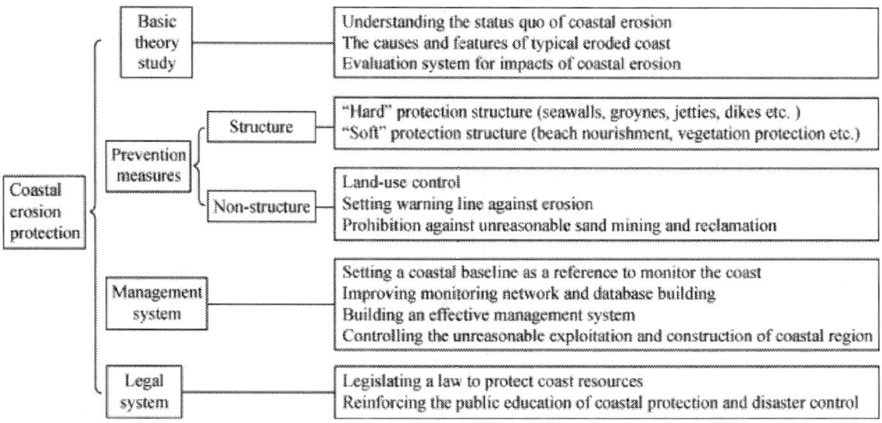

Figure 6. Sketch map of the prevention measures against the coastal erosion in China.

4.1. Basic research of coastal erosion and assessment of its impacts

According to studies since the early 1980s, the characteristics, processes and mechanisms of coastal erosion in different areas are basically understood and much valuable data have been obtained [70], [71], [72], [73], [74], [75], [76], [77] and [78]. A rapid change in the coastal environment has induced irreversible long-term erosion. To protect coastal property and reduce losses, a forward-looking analysis method of the erosion trend was constructed to provide a useful theoretical foundation for nationwide investigation and assessment [1], [79], [80], [81], [82], [83], [84], [85] and [86]. In recent years, the Chinese government has implemented the strategy of "Ocean Exploitation" and the policy of the "Development of National Ocean Economy". To meet the requirements of economic development and ocean management and comprehensively investigate marine resources and environmental conditions, in 2004, the state organized a large-scale marine survey called the "908 Special Project". Great progress in the basic theoretical study of coastal erosion is expected from the success of the "908 Special Project". For instance, a general understanding of the mechanisms, processes, and causes of nationwide current erosion, and knowledge of characteristics of seasonal and annual variations of typical coasts will provide fundamental data for disaster reduction, the rational use of coastal resources and assessment of the impacts of global climate change on coastal erosion. Scientific evaluation of the carrying capacity (including the carrying capacity of coastal resources), the coastal environment and the coastal ecological system that is affected by the coastal erosion will provide a scientific basis for changes to marine industries and the distribution of productivity, help plan the exploitation and economic development of coastal areas, promote the establishment of a "digital coast" and building of a fundamental database for coastal protection.

Monitoring the coast and beach erosion in different regions, determining its causes, mechanism, and trend, evaluating the impacts, and presenting protection measures are all necessary for the management and development of development projects. Coastal erosion is the main coastal geologic hazard, and like other hazards it has two types of attributes: natural and social attributes. From a comprehensive and correlative analysis of all attributes, further research will be conducted to promote quantitative and qualitative evaluations of the risks, erosion intensity, and financial losses during a coastal disaster, and to improve the forecasting of future erosion. In addition, new erosion type and intensity classifications, evaluation index systems and assessment models for the impacts of coastal erosion on the regional economy of surrounding areas, and erosion management and monitoring networks will be required.

In summary, the basic research mentioned above would provide a theoretical and scientific foundation for important decision making relating to the significant projects during exploitation, coastal zone management and environmental protection, and changes to the national economy development strategy.

4.2. Intensifying research on coastal protection measures and proper protection of typically eroded coast

Not many seawalls in China can stand the risk of being damaged by 1 in 100 year storm tides. A sea level rise caused by global warming will reduce a seawall's protective capability. Thus, determining how to reconstruct or reinforce seawalls and dikes is an urgent task.

There are two kinds of protective measures for controlling coastal erosion: structural measures and non-structural measures. Structures can be divided into hard and soft structures. Hard structures include seawalls, breakwaters, jetties, groins, and offshore dikes, whereas soft structures include beach nourishment, man-made dunes, planting of mangroves and rise grass, and coastal shelter belts. Non-structural measures include land-use controls, setting warning lines such as the coastal baseline and coastal construction control line to protect the coast from improper construction, and prohibition of unreasonable sand mining and reclamation.

Research and practices indicate the measures mentioned above have different scopes and prospects of application and economic and eco-environmental impacts (negative and positive) when put into practice. Therefore, according to the levels of economic development and coastal features in different regions, how to create proper protection measures against coastal erosion and determine an updated engineering standard are the major works for geomorphologists and coastal engineering scholars in China.

Principal protection measures before the 1960s were passive coastal defenses such as seawalls, breakwaters, and coastal shelter belts. Although they did reinforce the coast, they also caused much inconvenience during construction. Therefore, a new strategy of positive coastal protection has since then been put forward. In the new strategy, beach protection was put before coast protection and new kinds of groins, offshore dikes and their combinations were built in different regions. The aims of the strategy were to dissipate waves and tidal energy in front of the foreshore before they cause erosion, transform the previous integral protection into segment protection, reduce financial costs, and beautify the coastal environment. Over the past 10 years, a soft structural approach (such as beach nourishment and biological protection by mangrove planting), which is more natural and beautiful than building hard structures, has been greatly pursued [87], [88], [89] and [90].

With the development of tourism, the recreational function of beaches has attracted increasing attention. Thus, governments of many countries have put more effort into sandy beach protection. Beach nourishment, and mangrove and seaweed planting, which provide protection to backshore property and increase the recreational space along the shore, are popular in many developed countries [91], [92], [93], [94], [95] and [96]. In addition, the economic activity created by beach tourism has local benefits. To aid the coastal defense scheme, further scientific and technical research and benefit assessment are needed. It is suggested that beach nourishment should be a part of integrated coastal zone management.

4.3. Improvement of the integrated coastal zone management

Integrated coastal zone management (ICZM) is a dynamic, multidisciplinary and iterative process to promote sustainable management of the coastal zones. It covers the full cycle of information collection, planning (in its broadest sense), decision making, management and monitoring of implementation. ICZM uses the informed participation and cooperation of all stakeholders to assess the societal goals in a given coastal area, and to take actions towards meeting these objectives. ICZM seeks, over the long-term, to balance environmental, economic, social, cultural and recreational objectives, all within the limits set by natural dynamics. "Integrated" in ICZM refers to the integration of objectives and also to the integration of the many instruments needed to meet these objectives. It means integration of all relevant policy areas, sectors, and levels of administration, and integration of the terrestrial and marine components of the target territory, in both time and space (The Europe Commission).

The concept of ICZM was born in the mid-1980s, and its specific target was determined at the International Conference on Coastal Zone Management held in Noordwijk, The Netherlands, in 1993 [97]. China has implemented a new coastal zone management policy since the early

1990s to manage coastal resources and control environmental pollution.

Over the past 10 years, the State Oceanic Administration has conducted extensive studies including those on marine functional zoning, ocean space utilization and management, the promotion of legislation for marine regions delimiting, monitoring and evaluating aqueous environments, and established a series of coastal wetland reserves in regions where conditions permit, as well as presented research on a coastal zone management information system and a "digital ocean". Recently, technology integrating remote sensing, geographic information systems, global positioning systems and multiple sources of information has been applied in monitoring the temporal and spatial variations in coastal resources and environment [98], [99], [100], [101], [102] and [103]. Such technology would make a great contribution to ICZM and the sustainable exploitation and utilization of coastal resources. However, the ICZM project does not provide technical solutions to problems. In this section, pressing issues relating to the coastal protection in the new situation of climate change are presented.

1) China's coastal baseline should be indicated on large-scale charts to improve coastal monitoring and management. The baseline is a relatively fixed reference line determined by the changes in the long-term shoreline position, the regional coastal physical geography, and socio-economic development trends. It can be used as a technical reference for coastline monitoring and dynamic assessment, coastline management and implementation of coastal protection policy.
2) More efforts are needed in improving the nationwide dynamic monitoring network and database establishment program for coastal erosion, setting immovable monitoring profiles in eroded hotspots, and regular monitoring of the shoreline position, coastal topography and sediments. Aforementioned monitoring results are kept in a dynamic database so as to analyze the erosion trends promptly and provide fundamental data for coastal investigation, research and management.
3) Establishment of a nationwide shoreline management system requires the cooperation and coordination of different departments, support of research and monitoring results, public participation in decision making, and implementation of local government. Presently, governments of coastal provinces, cities, and autonomous regions have their own institutions for ocean

(aquatic) management with defective functions and regulations. There have been conflicts among various departments involved in shoreline and ocean management resulting in a waste of management resources. In summary, an unambiguous shoreline management plan is required.

4) It is necessary to stop excessive resource exploitation and unreasonable coastal construction. Unwise human activities are one of the most important factors causing coastal erosion, with human influences overwhelming natural influences. Shore sand mining is considered to have been the major cause of numerous cases of erosion, and unreasonable coastal engineering (e.g., tide gates, reclamation, and docks) also have negative effects on the environment. Thus, government should pay more attention to these problems.

4.4. New development of coastal protection measures and a management system

Coastal protection is an important part of territorial management. Facing serious coastal erosion aggravated by global warming, the following precautions should be taken. First, public awareness of coastal resources and environmental protection should be increased, and a new consciousness for disaster control planning needs to be popularized. Second, the health of the coastal environment and ecosystems needs to be maintained to guarantee the stability and security of the coast. In the future, based on countermeasure research and the integrated planning and supervision of sustainable coastal development, the government should draft an exploitation and utilization plan, make feasible policy for scientific management, and especially promote legislation to enhance coastal erosion control. Finally, administrative law enforcement should be strengthened for coastal protection.

ACKNOWLEDGEMENTS

This work was supported by the National Natural Science Foundation of China (Grant No. 40676051) and China 908 Special Project (908-02-03-04).

REFERENCES

1. Chen JY, Chen SL. Estuarine and coastal challenges in China. Mar Geol Lett 2002;18(1):1–5, [in Chinese].
2. Australian Bureau of Statistics. Regional population growth, Australia and New Zealand, 2001-02, Cat. No. 3218.0.
3. IPCC. Intergovernmental Panel on Climate Change. Climate change 1995. The science of climate change. Cambridge University Press; 1996. p. 572.
4. IPCC. Intergovernmental Panel on Climate Change. Climate change 1995. Impacts, adaptations and mitigation of climate change: scientific-technical analyses. Cambridge University Press; 1996. p. 878.
5. Devoy RJN. Implications of sea level rise for Ireland. In: Proceedings of the SURVAS expert workshop, ZMK University of Hamburg, 19–21 June 2000. p. 33–46.
6. Bird ECF. Coastline changes: a global review. Chichester: John Wiley-Interscience; 1985, p. 219.
7. Zhang K, Douglas BC, Leatherman SP. Global warming and coast erosion. Clim Change 2004;64(1/2):41–58.
8. Moore LJ, Benumof BT, Griggs GB. Coastal erosion hazards in Santa Cruz and San Diego counties, California. J Coastal Res 1999;28:121–39, [special issue].
9. Morton RA, McKenna KK. Analysis and projection of erosion hazard areas in Brazoria and Galveston counties, Texas. J Coastal Res 1999;28:106–20, [special issue].
10. Zhuang ZY. Severe shoreline retreat of some straight sandy coasts in Shandong peninsula. J Ocean Univ Qingdao 1989;19(1):90–7, [in Chinese].
11. Qian CL. Effects of the water conservancy projects in the Luanhe River basin on the Luanhe River delta. Acta Geogr Sin 1994;49(2):158–66, [in Chinese].
12. Ji ZX. The characteristics of coastal erosion and cause of erosion. J Nat Disaster 1996;5(2):65–75, [in Chinese].
13. Li CX, Feng Y. Coastal features and sea level rise in China. In: The impacts of sea level rise on Delta Regions in China and the

countermeasures. Beijing: Science Press; 1994. p. 29–39, [in Chinese].

14. Fan HQ, Liang SC. Study and management of mangrove in China. Beijing: Science Press; 1995, p. 189–202 [in Chinese].
15. Wang Y. Marine geography of China. Beijing: Science Press; 1996, p. 468–70 [in Chinese].
16. Wang J, Liu ZC. Possible response of hurricane frequency in northwest Pacific to global warming. Quaternary Sci 1991;3:277–80, [in Chinese].
17. Fan DD, Li CX. Complexities of Chinese coast in response to climate change. Adv Clim Change Res 2005;1(3):111–4, [in Chinese].
18. Yang GS, Shi YF. Advances in relative sea level rise and its impacts in the coast of China. Adv Earth Sci 1999;10(5):475–82, [in Chinese].
19. Ye DZ, Fu ZB. Major issues of global change science. Chin J Atmos Sci 1994;18(4):498–512, [in Chinese].
20. Zhu XD, Li YF, Gui F. The causes of coastal disasters and strategy for damage mitigation in China. J Nat Disasters 2001;10(4):26–9, [in Chinese].
21. IPCC Climate Change. The second scientific assessment report of the intergovernmental panel on climate change. London: Cambridge University Press; 1995. p. 359–406.
22. Burnn P. Sea level as a cause of shore erosion. J Waterways Harbours Division ASCE 1962(88):117–30.
23. Dean RG. Equilibrium beach profile: characteristics and applica- tions. J Coastal Res 1991;7:53–84.
24. Pilkey OH, Young RS, Riggs SR, et al. The concept of shoreface profile equilibrium: a critical review. J Coastal Res 1993;9:255–78.
25. Wang P, Davis RA. Depth of closure and the equilibrium profile—a case study from Sand Key, West-Central Florida. Shore Beach 1999;67:33–42.

26. Thieler ER, Pilkey OH, Yong RS, et al. The use of mathematical models to predict beach behavior for US Coastal Engineering: a critical review. J Coastal Res 2000;16(1):48–70.
27. Pilkey OH, Young RS, Bush DM. Comment on Sea level rise by Leatherman et al. EOS 2000;81:436.
28. Dubois RN. A re-evaluation of Bruun's Rule and supporting evidence. J Coastal Res 1992;8:618–27.
29. Stive MJF, de Vriend HJ. Modeling shoreface profile evolution. Mar Geol 1995;126:235–48.
30. Cowell PJ, Roy PS, Jones RA. Simulation of large-scale coastal change using a morphological behaviour model. Mar Geol 1995;126:46–61.
31. Andrew J, Cooper G, Pilkey OH. Sea level rise and shoreline retreat: time to abandon the Bruun Rule. Global Planet Change 2004;43:157–71.
32. Nicholls RJ. Coastal vulnerability assessment for sea level rise: evaluation and selection of methodologies for implementation. Technical report TR 98002. Caribbean Planning for Adaptation to Global Climate Change (CPACC) project; 1998.
33. Bray MJ, Hooke JM. Prediction of soft-cliff retreat with accelerating sea level rise. J Coastal Res 1997;13:453–67.
34. Dennis KC, Niang-Diop I, Nicholls RJ. Sea level rise and Argentina: potential impacts and consequences. J Coastal Res 1994;14:205–23, [special issue].
35. Volonte CR, Nicholls RJ. Sea level rise and Uruguay: potential impacts and responses. J Coastal Res 1994;14:262–84, [special issue].
36. Nicholls RJ, Leatherman SP, Dennis KC, et al. Impacts and responses to sea level rise: qualitative and quantitative assessments. J Coastal Res 1994;14:26–43, [special issue].
37. Cambers G. Planning for coastline change. Coastal development setback guidelines in Nevis. UNESCO/SeaGrant, University of Puerto Rico; 1998. p. 40.
38. UNESCO. Planning for coastline change. Guidelines for

construc- tion setbacks in the Eastern Caribbean islands. CSI info 4, UNESCO, Paris; 1997. p. 14.

39. Environment Waikato: Development setback lines for the Coro- mandel Peninsula. Environment Waikato: coastal hazards and development setback recommendations. Summary Report, May 2002.

40. Leatherman SP, Nicholls RJ, Dennis KC. Aerial videotape-assisted vulnerability analysis: a cost-effective approach to assess sea level rise impacts. J Coastal Res 1994;14:15–25, [special issue].

41. Silenzi S, Devoti S, Nisi MF, et al. Integrated hazard of Italian Coastal Plains with respect to relative sea level rise: a case study in Versilia Plain. Littoral 2002, The changing coast. EUROCOAST/ EUCC, Porto-Portugal. Ed. EUROCOAST, Portugal; 2002. p. 115– 24.

42. Li CX, Fan DD, Deng B, et al. The coasts of China, and issues of sea level rise. J Coastal Res 2004;51(43):36–49.

43. Ren ME. The trends of sea level rise in Yellow River, Yangtze River and Pearl River deltas and the forecast of sea level rise in 2050. In: The impacts of sea level rise on Delta Regions in China and the countermeasures. Beijing: Science Press; 1994. p. 18–28, [in Chinese].

44. The State Oceanic Administration. China Sea Level Bulletin 2000; 2004 [in Chinese].

45. The State Oceanic Administration. China Sea Level Bulletin 2006; 2007 [in Chinese].

46. Yang GS. The regional response to changing coastal environment in China. Beijing: Higher Education Press; 2002, p. 60–1 [in Chinese].

47. Bird ECF. Shoreline changes during the past century. In: Proceedings of the 23rd international geographical congress. Moscow: IGC; 1976.

48. Sun Q, Zhang YS, Hu EH. Potential impacts of sea level rise on the economy and environment in the Yangtze River delta and the countermeasures thereof. Resour Environ Yangtze Basin

1997;6(1):58–64, [in Chinese].

49. Zuo SH, Li JF, Chen SL, et al. Study of delta coastal erosion and protection engineering measures in China—taking the Yellow river delta and the Yangtze river delta as examples. Chin J Geol Hazard Control 2006;17(4):97–101, [in Chinese].
50. Zuo SH, Li JF, Chen SL. The causes of coastal erosion and protection engineering measures in China. Yellow River 2006;28(1):23–6, [in Chinese].
51. Li HP, Yang GS. Determination of classification and risk in the coastal zone of Yangtze River delta and north Jiangsu. J Nat Disasters 2001;10(4):20–5, [in Chinese].
52. Zhu JW, Ji ZX, Jiang ZX. Impacts of sea level rise on the Changjiang River delta and its adjacent area. Sientia Geogr Sin 1994;14(2):109–17, [in Chinese].
53. Ke C. Study on the erosion and its countermeasures of the modern Yellow River Delta Coast. Master thesis, Ocean University of China, Qingdao, China, 2004 [in Chinese].
54. Chen SL, Zhang GA, Gu GC. Mechanism of heavy coastal erosion on Yellow River delta and its countermeasures. J Hydraul Eng 2004;7:1–7, [in Chinese].
55. Chen XY, Chen SL, Yu HJ, et al. Coastal profile types and evolution regularities of the Yellow River delta. Adv Mar Sci 2005;23(4):438–45, [in Chinese].
56. Sun YF, Duan Y, Wu SY, et al. Coastal evolution in the north of the Yellow River delta. Mar Geol Lett 2006;22(8):7–11, [in Chinese].
57. Wang F, Tian SZ. The impacts of sea level rise on the socio-economic and environment of Pearl River delta. Disaster Reduction China 2000;10(2):33–7, [in Chinese].
58. Li XJ. Application of the remote sensing method in the analysis of the shoreline change and its environmental impact in the Lingdingyang bay, Pearl River estuary, Guangdong, China. Geol Bull China 2007;26(2):215–22, [in Chinese].
59. Hsu TW, Lin TY, Tseng IF. Human impact on coastal erosion in Taiwan. J Coastal Res 2007;23(4):961–73.

60. Chen WJ, Kuo CT. Sea level variation around Taiwan. Available from: http://kfki.baw.de/fileadmin/conferences/ICHE/2000-Seoul/ pdf/51/PAP_098.PDF [23.03.08].
61. Chen WJ, Kuo CT. Shoreline changes of Tainan coast in Taiwan. Available from: http://kfki.baw.de/fileadmin/conferences/ICHE/2002-Warsaw/ARTICLES/PDF/47E1.pdf [23.03.08].
62. Emanuel KA. The dependence of hurricane intensity on climate. Nature 1987;325:483–5.
63. Sun JR, Duan YH. A study of the relationships between the East Asian summer monsoon and the tropical cyclone frequency in the northwestern pacific. Chin J Atmos Sci 2003;27(1):67–74, [in Chinese].
64. Zhang QY, Peng JB. The interannual and interdecadal variations of East Asian summer circulation and its impact on the landing typhoon frequency over China during summer. J Atmos Sci 2003;27(1):97–106, [in Chinese].
65. State Oceanic Administration. China Marine Disaster Bulletin; 2005.
66. Cai F, Su XZ, Xia DX. Study on the difference between storm effects of beaches on two sides of the tropical cyclone track—taking the response of beaches to No. 0307 typhoon Imbudo as an example. Adv Mar Sci 2004;22(4):436–45, [in Chinese].
67. Cai F, Su XZ, Yang SL, et al. A rapid response to 9914 typhoon-induced storm. Ocean Eng 2002;20(2):86–90, [in Chinese].
68. Cai F, Lei G, Su XZ, et al. Study on process response of Fujian beach geomorphology to typhoon Aere. Ocean Eng 2006;24(1):98–109, [in Chinese].
69. Yi YH, Chen ZX. Sea level change and the influences on coastal zone. Mar Geol Lett 2003;19(1):14–7, [in Chinese].
70. Wang WH. Mechanism and the countermeasures of coastal erosion in China. Ocean Exploitation 1987;1:8–12, [in Chinese].
71. Wang Y, Wu XG. Sea level rise and beach erosion. Acta Geogr Sin 1995;50(2):118–27, [in Chinese].
72. Zhang YH. Disaster of coastal erosion and prevention measures in China. Catastrophology 1996;11(3):15–21, [in Chinese].

73. Ruan CJ, Xie QL, Xu J. Coastal erosion and protection measures in China. J Soil Water Conserv 2000;14(1):44–7, [in Chinese].
74. Sheng JF, Zhu DK. Discussion about coastline erosion and management. Mar Sci Bull 2002;21(4):50–7, [in Chinese].
75. Wang YG, Zhang XW, Jia K, et al. Study on erosion along the Suizhong coast of Liaodong bay. Coastal Eng 2007;26(1):1–5, [in Chinese].
76. Li XN. Study on coastal erosion and accretion in Tianjin. Master thesis, China University of geosciences, Beijing, China; 2007 [in Chinese].
77. Chen XY, Chen SL, Li JF, et al. Study on the coastal erosion mechanism in the Yellow River delta Gudong and Xintan areas. Coastal Eng 2005;24(4):1–10, [in Chinese].
78. Feng AP, Xia DX, Gu DQ, et al. Study on process and cause of the coastal erosion along the south coast of the Laizhou Bay. Adv Mar Sci 2006;24(1):83–90, [in Chinese].
79. He QX. Challenges to the integrated coastal zone management in China. Mar Geol Lett 2002;18(4):1–5, [in Chinese].
80. Shi YF, Zhu JW, Xie ZR, et al. Forecast of the influences of sea level rise on Yangtze River delta and its adjacent areas and countermea- sures. Sci China Ser D 2000;30(3):225–32, [in Chinese].
81. Li JL, Yang XP, Tong YQ. Progress on environmental effects of tidal flat reclamation. Prog Geogr 2007;26(2):43–51, [in Chinese].
82. Ji ZX, Jiang ZZ, Zhu JW, et al. Possible influence of sea level rise on the coastal erosion of Yangtze River delta and south Jiangsu province. Acta Geogr Sin 1993;48(6):516–26, [in Chinese].
83. Zhang RS, Lu LY, Wang YH. The mechanism and trend of coastal erosion of Jiangsu Province in China. Geogr Res 2002;21(4):469–78, [in Chinese].
84. [84] Wu RJ, Zheng YF, Zhu MY, et al. Creeping environmental problems in coastal zone in China. J Mar Sci 2007;25(4):66–73, [in Chinese].

85. Yi MQ, Li C, Han SP, et al. Discussion on disaster forecast of erosion power at Yellow River coast. Res Soil Water Conserv 2007;14(1):315–8, [in Chinese].

86. Liu DJ. Possible influences of relative sea level rise on coastal zone in China. Mar Forecasts 2004;21(2):21–8, [in Chinese].

87. Xie SL. Beach improvement works at Guilinyang. Port Eng Technol 1993;1:1–8, [in Chinese].

88. Ji XM, Zhang YZ, Zhu DK. Evolution of Sanya coast and artificial beach design. Quaternary Sci 2007;27(5):853–60, [in Chinese].

89. Song XQ, Guo ZJ, Chen SM. The planning and design of artificial beach in the Xinghai Bay. China Civil Eng J 2005;38(4):134–40, [in Chinese].

90. Chen J, Cai F, Xu J, et al. Study on beach filling and reconstruction on northeast coast of Xiamen island. J Oceanogr Taiwan Strait 2002;21(2):243–51, [in Chinese]. Benassai E, Gentilomo M, Ragone A, et al. Littoral restoration by means of protected beach nourishment—recent Italian works. PIANC Bull 1997;94:43–55.

91. Dette HH. Beach fill practice in Germany. SAFE report, 2nd annual workshop. Commission of the European Communities, Sylt, Ger- many, 1998.

92. Hamm L, Capobianco M, Dette HH, et al. A summary of European experience with Shore nourishment. Coastal Eng 2002;47:237–64.

93. Houston JR. The economic value of beaches—a 2002 update. Shore Beach 2002;70(1):9–12.

94. Micallef A, Williams AT, Cassar M. Environmental risk assessment: application to a proposed beach nourishment, Malta. Shore Beach 2001;69(3):13–7.

95. Benassai E, Calabrese M, Uberti GSD. A probabilistic prediction of beach nourishment evolution. In: Medcoast 01: proceedings of the fifth international conference on the mediterranean coastal environ- ment, vol. 1. Berlin: Springer; 2001. p. 1323–32.

96. World Bank. Guidelines for integrated coastal zone management. Issued at the World coast conference. Noordwijk, The Netherlands; 1993.

97. Huang JL, Hong HS, Zhang LP. Application of geo-informatics in resources and environment management in coastal zone. J Oceanogr Taiwan Strait 2003;22(1):79–84, [in Chinese].

98. Jiang YW, Hong HS, Zhang LP. The application of geographic information system (GIS) in the marine water quality numerical model. J Xiamen Univ (Nat Sci) 1999;38(1):90–5, [in Chinese].

99. Chang J, Liu GH, Liu QS. Dynamic monitoring of coastline in the Yellow River delta by remote sensing. Geo-Inf Sci 2004;6(1):93–8, [in Chinese].

100. Wang L, Xu HQ, Li S. Dynamic monitoring of shoreline changes in Xiamen island with its surrounding areas of SE China using remote sensing technology. Remote Sens Technol Appl 2005;20(4):404–10, [in Chinese].

101. [Budetta P, Santo A, Vivenzio F. Landslide hazard mapping along the coastline of the Cilento region (Italy) by means of a GIS-based parameter rating approach. Geomorphology 2008;94: 340–52.

102. Jacek AU, Gabriela G, Natalia C, et al. A GIS-based WFD oriented typology of shallow micro-tidal soft bottom using wave exposure and turbidity mapping. Estuarine Coastal Shelf Sci 2008;78(1):1–11.

Chapter 8

The Determination of Sea Tourism Season with Respect to Climatical Conditions on the Black Sea Region of Turkey

Yuksel Guclu*

Sakarya University, Faculty of Education, Department of Social Studies Education, 54300, Hendek / Sakarya, Turkey

ABSTRACT

The aim of the study is to determine the favourable season according to the climatical condition in respect to sea tourism season on the Black Sea Region coastal belt of Turkey. The climatical elements bearing importance for the sun and the sea bathing; sunshine duration, temperature, relative humidity, wind, precipitation and sea surface temperature have been taken into consideration for this purpose. Climatic conditions in respect to sunbathing and sea bathing in coastal belt between Akçakoca and Samsun are favourable in general. But, climatic conditions in respect to sunbathing and sea bathing in coastal belt between Samsun and Hopa are not favourable in general. High rainfall and relative humidity, short sunshine duration and low temperature are the most important climatic problems in the study area in respect of sea tourism.

Keywords: Climate; Sea tourism; Tourism season; Black Sea

INTRODUCTION

"Tourism is one of the world's largest, fastest growing and most climate-dependent economic sectors" [1]. The tourist attraction potential of any place on Earth depends on the attractions, access and accommodation of that spot. These are also called three S of the tourism [2].

The attractions create the desire to visit a spot and also are geographical sources that play role in the place selection of the tourism investments. Since nearly the preferences and expectations of each individual are different from each other, it is not easy to determine which geographical characteristics attract tourists. However, climatic features are certainly important among such attractions [2]. According to de Freitas (2003), the climate is one of the natural sources of tourism like geographical location, orography, geographical formation etc. [3].

The climate provides an advantage to tourism areas; however, it sometimes brings disadvantages [3]. For instance, dry and sunny summers provide important advantages for the sea tourism in locations where the sea- sun-sand trio demonstrates desirable combinations. The rainy summers affect adversely the sea tourism and weak snowy winters affect the skiing tourism as well [4].

When we examine the tourism areas and tourism movements in the world, the first point worth attention is that the tourism has a seasonal character. Whereas some tourism activities are intense only in a certain season, some demonstrate variations according to the tourism season. In addition, the travels of the visitors to certain areas vary depending on the season [1]. Thus, the tourism activities happen to be intense in certain periods all over the globe. The length of the season in which tourism may take place changes according to the dominant climate conditions in that area.

A big portion of world population perceives the tourism as going to seaside for vocation. Owing to this reason, the most popular tourism areas are seaside locations [2]. The coastal belts are suitable places for swimming, sea bathing, observation of underwater life, wreck seeing, diving, yachting, fishing, sunbathing, several water sports (water skiing etc.), and sightseeing etc. recreational activities. However, the tourists preferring the coastal belts in the first place desire the beaches suitable for swimming and sunbathing [2].

The coasts in Turkey are the most important holiday locations. Both the domestic and international tourism concentrate on the coastal belt and in the summer season [2]. When considered the development process of the tourism and development areas of the same, such situation is quite eminent. The tourism in Turkey depends predominantly on the sea-sun-sand trio (also known as three S of tourism) and accordingly has flourished and developed on the coastal regions.

The variety in the physical features of the coastal areas determines the tourism usages of the shores in Turkey. For instance, Western Mediterranean and Southern Aegean coasts support international tourism depending on the basis of the sea-sun-sand concept; Black Sea and Marmara coasts attract recreational activities in differing importance and density with domestic tourism [2].

Since sea tourism is a type depending on different recreational activities related to the sea, the physical geography features affective in the development of sea tourism may be grouped to three as the geomorphologic characteristics of the coasts, climate conditions and flora [5]. The climate is the most important factor in sea tourism, as many other tourism types. The sea tourism is more intense in the summer due to such climate relation and its length is determined by the air and sea temperatures [5].

This study aims to determine the sea tourism season of the Black Sea region of Turkey in relation with climatic conditions. This is because the required respective information shall be established for favouring the sea tourism under healthy and comfortable conditions, obtaining the expected favors from the tourism, protecting the tourism activity participators from the possible climate risks, and enabling the tourism planners and marketers to perform in realistic manners. Moreover, determination of the sea tourism season is thought to contribute to the real potential of the Black Sea Region coastal belt. Besides, frequent views are being expressed claiming that the climate changes would make Mediterranean and Aegean coats less attractive in future by means of sea tourism. Even now, in the Mediterranean and Aegean coasts, sea tourism experiences climatic problems where in July and August, the air and sea temperatures demonstrate high values from time to time. Thus, it is being thought that the Marmara and, particularly, Black Sea regions will be preferred in future.

STUDY AREA

The study area comprises the coastal belt between Akçakoca and Hopa in the Black Sea Region of Turkey (Fig. 1). The most known centres of sea tourism in the Black Sea Region coastal belt are Akçakoca, Amasra, Kurucaúile, Cide, Abana, ønebolu, Sinop, Gerze and Samsun. However, particularly compared to the coastal belts of Mediterranean, Aegean and, partly, Marmara with Black Sea, it can be seen that sea tourism is not sufficiently favored in the latter [6]. The shores of the Black Sea Region coastal belt, where the sea tourism potential is high include sea tourism activities which are being favored mostly by the local habitants and certain amount of domestic tourists arriving from centres like Istanbul and Ankara.

Figure 1. Location map of the study area

Since the features of the coasts are important for the sea tourism, we shall, in short, define the general characteristics of the Black Sea Region coastal belt. The Black Sea Region coasts proceed longitudinally; i.e. "Pacific coastal type". These coasts do not contain many protruding- indented coasts and certain amount of sea cliffs are worth noting. This situation is the result of the geological and geomorphologic evolution experienced by the region. Whereas the Northern Anatolian Mountains, within the Alpine orogenic belt, have risen since Oligocene era, the Black Sea basin has been deepened. This epirogenic movement is repeated in Quaternary era and the slope in the coastal area has increased and the continental shelf has been bent and narrowed [7]. Tectonic activities, on the other hand, have caused curling and faulting in the Black Sea coastal area. Moreover, the eustatic level changes in Quaternary have partially affected the shaping of the shores and delta formation.

The portion between Ünye and Sarp of the Black Sea Region coastal belt is high and has a sea cliff structure. In the middle Black Sea coasts, Kızılırmak and Yeúilırmak deltas are present with a coastal area including lower beaches. The traces of younger tectonic movements may be seen around the Sinop peninsula. The curved structure of Ere÷li and around and experienced tectonic movements have caused wavy formations [7]. The beach line, which carries utmost importance for sea tourism, is very rare in the Black Sea coastal belt due to the geological and geomorphologic features of the same. The rarity of the bays that enter deeply towards mainland and lower shores, deepening of the sea in short distances have limited the development of sea tourism [5]. The shores demonstrating spread beach lines may be found in Kurucaúile and Cide. A thin coastal line present in between ønebolu and Abana [6].

MATERIAL AND METHOD

The study has used the sunshine duration, temperature, relative humidity, precipitation, wind speed and sea surface temperature data of 11 stations (between Akçakoca and Hopa) in the Black Sea Region coastal belt through years 1975 and 2008. The selection of the stations depended mostly on more than 25 years of observation periods of the stations and their ability to measure the sea surface temperature. These data have been obtained from the Ministry of Environment and Forestry, Turkish State Meteorological Service.

The climate data of the meteorology stations selected for this study have been examined by referencing the threshold data established in this branch and the obtained results have been transferred to the tables and interpreted. Below are the generally accepted climate factors and assessments related to the threshold values with effect to the sea tourism.

Sunshine condition: It is important for the weather to be open or slightly cloudy and without rain, reaching of the sun rays to earth with appropriate angles and the temperature to be in appropriate values together with the relative humidity and wind in respect of sun bathing [2].

Temperature: "According to Maddison (2001) the maximum daytime temperature should ideally be close to 30ºC, while Lise and Tol (2002) estimate the optimal 24-hour mean daily temperature to be around 21ºC." [8]. "Globally, OECD tourists prefer a temperature of 21°C (average of the hottest month of the year) in their holiday destination choice" [9].

According to Özgüç [2], the temperatures in which the humans may live with light clothes are the temperatures over 16 °C-18 °C. On the other hand, the air temperatures at least 20°C, as a normal value 25- 28°C, at most 32-34°C, as for the sea surface temperatures 22-25°C, as an extremely value 18-28°C are accepted as the suitable values [10].

Wind speed: The wind is an important climate factor for many activities, such as swimming, sunbathing, windsurf etc., in sea tourism coverage. On the other hand, it is an important factor for thermal comfort. It affects the feeling of temperature and makes people feel the air temperature higher or lower [11, 12].

For the recreational activities, the wind speed in the atmosphere shall not exceed 7 m/sec [5]. According to the health climatology experts, the mean wind speed being lesser than 6 m/sec is positive for climate comfort requirements [13]. A wind blowing in light intensity over the sea is considered suitable for recreational activities such as windsurfing, sailing etc. The mild wind blowing from the sea makes it easy to endure the heat through removing moisture from the skin. The main reason for tourists preferring daily boat trips in holiday resorts is that to favor the fresh effects of the winds on the sea to save them from the suffocating hotness of summer. Especially, land and sea breezes are important in this context. For example, the conversion time of boat tours is accordingly adjusted [5].

Precipitation: Precipitation characteristics are important for tourism activities particularly for the period corresponding to tourism season. Rainless weather conditions are important for activities like sun bathing, benefiting from sea, visiting tourism-wise attractive areas [2]. Therefore, rainless weather is very important for the marine tourism especially in summer [5].

Relative humidity: Relative humidity is higher in sea costs and in areas dominated by marine influence. According to studies, the excess of the relative humidity in air increases the effects of the air temperature on human [5]. Thus, when high temperatures and relative humidity co-exist, annoyance is the case and health problems may occur. And this may be a climate barrier preventing the development of tourism [2]. On the other hand, the high relative humidity will produce a foggy and misty environment and weaken the sun lights and decreases the sunbathing comfort. On the contrary, if the relative humidity is low, i.e. the air is dry, the sunbathing comfort will improve.

Sea surface temperature: For the sea bathing, besides the cleanliness of the sea water, the ideal sea surface temperatures are accepted as 22-25°C and the acceptable threshold values are regarded as 18-28 °C [10].

RESULTS

Sunshine condition

As selected stations, along the Black Sea Region coastal belt, sunshine reaches the surface with angles from 25█2' to 70⍰57' within the year. These angles vary between 49° to 71°, when sea tourism activities become intense (Table 1). The real sunshine durations in annual average change between 4, 09 hours (Rize) and 5, 58 hours (Zonguldak). The mean annual theoretical sunshine durations are around 12 hours, and the mean annual sunshine rates change between 34% - 48, 3% and not exceed 50% (Table 1).

Table 1. Sunshine conditions of selected stations along the Black Sea Region coastal belt of Turkey (1975-2008 period)

Stations	Meteorological elements	Months												Yearly mean
		I	II	III	IV	V	VI	VII	VIII	IX	X	XI	XII	
Akçakoca Lat.(41°05′ N)	Theor.sun.dur. (Hour.min.)	9,35	10,38	12,00	13,20	14,26	15,08	14,51	13,51	12,40	11,12	9,56	9,09	12,03
	Real sun. dur. (Hour/min.)	2,12	2,54	4,06	4,54	6,42	8,36	9,30	9,06	7,06	4,54	3,00	2,00	5,25
	Sun. ratio (%)	22,7	24,5	33,8	34,4	45,0	55,4	64,0	67,1	56,9	40,8	31,4	22,0	43,6
	Declination of Sun. rays	32° 12′	39° 56′	47° 40′	55° 19′	63° 13′	70° 53′	66° 10′	58° 26′	49° 57′	42° 53′	34° 59′	26° 49′	
Zonguldak Lat.(41°27′ N)	Theor.sun.dur. (Hour:min.)	9,33	10,37	12,00	13,22	14,29	15,12	14,54	13,54	12,40	11,10	9,54	9,06	12,00
	Real sun. dur. (Hour/min.)	2,24	3,06	4,30	5,42	7,48	9,36	10,24	9,36	7,42	5,12	3,18	2,18	5,58
	Sun. ratio (%)	24,0	29,5	35,8	41,0	52,3	61,9	70,4	69,1	59,8	46,1	33,3	24,1	46,5
	Declination of Sun. rays	31° 48′	39° 32′	47° 16′	54° 55′	62° 49′	70° 28′	65° 46′	58° 02′	50° 33′	42° 29′	34° 35′	26° 25′	
Amasra Lat.(41°45′ N)	Theor.sun.dur. (Hour:min.)	9,32	10,37	12,00	13,23	14,30	15,13	14,55	13,55	12,40	11,10	9,53	9,05	12,04
	Real sun. dur. (Hour/min.)	2,42	3,18	4,24	5,18	7,24	8,48	10,00	9,24	7,24	5,48	3,48	2,42	5,50
	Sun. ratio (%)	26,0	30,1	35,3	39,2	50,6	56,0	68,7	68,2	58,4	49,4	36,5	26,7	45,7
	Declination of Sun. rays	31°37′	39° 21′	47° 05′	54° 44′	62° 38′	70° 17′	65° 35′	57° 51′	50° 22′	42° 18′	34° 24′	26° 14′	
İnebolu Lat.(41°59′ N)	Theor.sun.dur. (Hour:min.)	9,31	10,36	12,00	13,24	14,31	15,15	14,56	13,56	12,40	11,10	9,53	9,04	12,00
	Real sun. dur. (Hour/min.)	2,18	3,00	4,00	4,54	6,54	9,06	10,00	8,54	6,48	4,42	3,24	2,18	5,32
	Sun. ratio (%)	23,4	29,0	33,3	34,3	45,7	59,8	68,7	63,0	52,3	39,8	34,0	24.1	44,3
	Declination of Sun. rays	31°16′	39°00′	46° 44′	54°23′	62° 17′	69° 56′	65° 16′	57° 30′	50° 01′	41° 57′	34°03′	25° 53′	

Sinop Lat.(42°03′ N)	Theor.sun.dur. (Hour:min.)	9,28	10,35	12,00	13,27	14,34	15,19	14,59	13,59	12,40	11,08	9,51	9,01	12,05
	Real sun. dur. (Hour/min.)	2,18	3,18	4,12	5,12	6,48	8,42	9,30	8,36	6,42	4,48	3,24	2,06	5,28
	Sun. ratio (%)	23,5	30,7	34,3	38,6	45,2	55,4	63,7	61,5	51,8	40,4	34,1	22,9	43,8
	Declination of Sun. rays	30° 55′	38° 39′	46° 23′	54° 02′	61° 55′	69° 35′	64° 59′	57° 09′	49° 40′	41° 36′	33° 42′	25° 32′	
Samsun Lat.(41°21′ N)	Theor.sun.dur. (Hour:min.)	9,34	10,37	12,00	13,22	14,29	15,12	14,54	13,54	12,40	11,10	9,54	9,06	12,04
	Real sun. dur. (Hour/min.)	3,00	3,36	4,12	5,12	6,36	8,42	9,24	8,48	6,54	5,00	4,12	2,54	5,38
	Sun. ratio (%)	32,1	32,4	34,3	38,7	44,5	55,7	63,5	62,6	52,7	45,0	43,2	28,0	44,7
	Declination of sun. rays	31° 58′	39° 42′	47° 26′	55° 05′	62° 59′	70° 39′	65° 56′	58° 12′	50° 43′	42° 39′	34° 45′	26° 35′	
Ordu Lat.(40°59′ N)	Theor.sun.dur. (Hour:min.)	9,35	10,38	12,0	13,20	14,26	15,08	14,51	13,51	12,40	11,12	9,56	9,09	12,03
	Real sun. dur. (Hour/min.)	2,24	2,54	3,24	4,06	5,42	6,48	6,18	6,12	5,12	4,06	3,24	2,12	4,23
	Sun. ratio (%)	23,9	24,5	27,0	30,8	38,0	43,0	42,6	45,3	41,3	36,5	33,9	23,3	35,1
	Declination of Sun. rays	32° 16′	40° 00′	47° 44′	55° 23′	63° 17′	70° 57′	66° 14′	58° 30′	51° 01′	42° 57′	35° 03′	26° 53′	
Trabzon Lat.(41°00′ N)	Theor.sun.dur. (Hour:min.)	9,35	10,38	12,00	13,20	14,26	15,08	14,51	13,51	12,40	11,12	9,56	9,09	12,03
	Real sun. dur. (Hour/min.)	2,24	2,54	3,18	4,12	5,42	6,36	5,54	5,12	4,54	4,24	3,24	2,18	4,16
	Sun. ratio (%)	24,0	24,5	26,5	31,2	38,0	42,2	38,2	37,9	36,6	38,1	33,9	24,0	34,6
	Declination of Sun. rays	32° 15′	39° 59′	47° 43′	55° 22′	63° 16′	70° 56′	66° 13′	58° 29′	51° 00′	42° 56′	35° 02′	26° 52′	

Rize Lat.(41°02′ N)	Theor.sun.dur. (Hour:min.)	9,35	10,38	12,00	13,20	14,26	15,08	14,51	13,51	12,40	11,12	9,56	9,09	12,03
	Real sun. dur. (Hour/min.)	2,06	2,48	3,30	4,30	5,54	6,18	5,24	5,06	5,06	4,06	2,54	2,06	4,39
	Sun. ratio (%)	22,0	23,9	27,5	32,6	38,8	41,0	36,1	37,5	40,8	36,5	26,6	22,7	36,5
	Declination of sun. rays	32° 13′	39° 57′	47° 41′	55° 20′	63° 14′	70° 54′	66° 11′	58° 27′	49° 58′	42° 54′	35° 00′	26° 50′	
Hopa Lat.(41°24′ N)	Theor.sun.dur. (Hour:min.)	9,33	10,37	12,00	13,22	14,28	15,11	14,53	13,53	12,40	11,11	9,55	9,07	12,04
	Real sun. dur. (Hour/min.)	2,24	3,00	3,54	4,42	5,54	6,00	5,24	5,18	5,42	4,36	3,42	2,36	4,26
	Sun. ratio (%)	24,0	28,9	29,5	33,4	38,8	39,7	36,1	38,3	43,7	39,2	35,8	26,0	35,4
	Declination of sun. rays	31° 51′	39° 35′	47° 19′	54° 58′	62° 52′	70° 32′	65° 49′	58° 05′	50° 36′	42° 32′	34° 38′	26° 28′	

Abbreviations: Theor.sun.dur.: Theoretical sunshine duration; Real sun.dur.: Real sunshine duration; Sun. ratio: Sunshine ratio; sun. rays: sunshine rays; min: minute; Lat.: Latitude. Source: Theoretical sunshine duration is calculated according to [14], page 12, figure 4. The sunshine durations are organized according to data of the Ministry of Environment and Forestry, Turkish State Meteorological Service. Sunshine ratios and declinations of sunshine rays are calculated.

When we consider the five month period as May-September in which sea tourism activities are intense, the mean monthly real sunshine durations are from 4,5 hours (Trabzon-September) to 10,24 hours (Zonguldak: July). The longest periods in Akçakoca-Ordu coastal belt are spotted in July and in Ordu-Hopa coastal belt in June. The daily real sunshine durations, in the months suitable for sea tourism, vary from 5, 06 hours (Rize) to 12, 18 hours (Zonguldak). The sunshine rates in this season also vary from 36, 1% (Rize and Hopa) to 70, 4% (Zonguldak) (Table 1). The annual total sunshine durations change between 1519 - 2187 hours.

The cloudiness rate, an important factor for insolation, varies about 5, 5/ 10-6, 5/ 10 throughout the year. Particularly in the Eastern Black Sea Sub region coastal belt, it is higher than 6, 0/ 10. It lessens in Western and Middle Black Sea; however, still close to 6, 0/ 10 [15]. It decreases in the summer and varies as 5/ 10-6/ 10 and as 3/ 10-4/ 10 in Western and Middle Black Sea coastal belt [6].

The sunshine durations and rates in the Black Sea Region coastal belt are not as long as the ones in the coasts of Aegean and Mediterranean Regions. The most suitable conditions in the Black Sea Region coastal belt for insolation are observed between mid June and mid August. However, a suitable period for insolation is not present for the line between Samsun and Hopa because sunshine durations decrease. The daily sunshine durations do not reach to 10 hours and above. The coasts between Akçakoca and Samsun have the best insolation conditions.

Temperature

With respect to the selected stations, the mean annual temperatures change between 12, 6°C (Akçakoca) - 14, 5°C (Trabzon) along the Black Sea Region coastal belt of Turkey (Table 2). Two months per year for Akçakoca-Sinop coastal belt (July and August), three months for Samsun, Ordu, Rize and Hopa (summer months), four months for Trabzon and Giresun (June to September) have mean temperatures over 20°C (Table 2).

Table 2. The mean monthly, annual and certain period temperature values of the stations selected from the Black Sea Region coastal belt of Turkey (1975-2008 period)

Meteorological stations	Mean monthly temperature values (°C)												Mean annual temper. values (°C)	Mean temper. values in May-September period (°C)	Mean temper. values in June-August period (°C)
	I	II	III	IV	V	VI	VII	VIII	IX	X	XI	XII			
Akçakoca	5,2	5,0	6,6	10,3	14,5	18,9	21,3	21,2	17,6	13,9	9,7	7,1	12,6	18,7	20,5
Zonguldak	6,0	5,7	7,4	11,3	15,2	19,5	21,7	21,7	18,5	14,9	11,1	8,0	13,4	19,3	21,0
Amasra	6,2	5,8	7,3	10,8	14,8	19,3	21,7	22,0	19,0	15,3	11,4	8,2	13,5	19,4	21,0
İnebolu	5,7	5,3	6,8	10,4	14,5	19,1	21,7	21,6	18,1	14,3	10,5	7,5	13,0	19,0	20,8
Sinop	6,8	6,2	7,4	10,6	14,7	19,6	22,6	23,0	19,7	15,9	11,9	8,8	13,9	19,9	21,7
Samsun	7,0	6,5	7,9	11,2	15,3	20,1	23,2	23,5	19,9	15,9	11,8	8,8	14,3	20,4	22,3
Ordu	6,7	6,4	7,9	11,4	15,5	20,2	23,0	23,3	19,9	15,8	11,5	8,5	14,2	20,4	22,2
Giresun	7,2	6,7	8,1	11,5	15,3	20,0	22,9	23,2	20,0	16,2	12,2	9,1	14,4	20,3	22,0
Trabzon	7,3	6,9	8,3	11,9	15,7	20,2	23,0	23,1	20,0	16,2	12,2	9,2	14,5	20,4	22,1
Rize	6,4	6,1	7,9	11,8	15,9	20,3	22,8	23,1	19,9	15,9	11,4	8,1	14,1	20,4	22,1
Hopa	6,9	6,8	8,5	12,2	15,8	20,0	22,5	22,8	19,4	15,5	11,6	8,8	14,2	20,1	21,8
Mean	*6,5*	*6,1*	*7,6*	*11,2*	*15,2*	*19,7*	*22,4*	*22,6*	*19,3*	*15,4*	*11,4*	*8,4*	*13,8*	*19,8*	*21,6*

Abbreviations: temper: temperature. Source: Calculated and organized according to the data of the Ministry of Environment and Forestry, Turkish State Meteorological Service.

Table 3. Favourable air temperature periods for sea bathing of selected stations along the Black Sea Region coastal belt of Turkey (1975- 2008 period)

Meteorological stations	Favourable air temperature periods for sea bathing (20–32 °C) Starting and termination dates	Number of days
Akçakoca	24 June - 29 August	67
Zonguldak	13 June - 2 September	82
Amasra	13 June -30 August	79
İnebolu	23 June - 2 September	72
Sinop	14 June - 18 September	97
Samsun	12 June - 16 September	97
Ordu	13 June- 16 September	96
Giresun	13 June- 16 September	96
Trabzon	9 June- 16 September	100
Rize	9 June- 16 September	100
Hopa	13 June- 8 September	88
Mean	*14 June-10 September*	*89*

Source: Calculated and organized according to the data of the Ministry of Environment and Forestry, Turkish State Meteorological Service.

Considering the mean temperature of the hottest month, it may be seen that the hottest month is August (except for Amasra), July in between Akçakoca and Sinop (July and August values are the same for Zonguldak) and August between Sinop and Hopa (Table 2). The mean daily temperatures in the Black Sea Region coastal belt, in the high tourism season, may not exceed 24, 2°C where the highest temperatures also may not exceed 28, 2°C.

The ideal air temperature threshold values for sea bathing are accepted as 25-28°C; a certain suitable period may not be named since the average temperatures do not exceed 25°C throughout the year in the stations selected from Black Sea Region coastal belt. When the suitable air temperature threshold values for sea bathing are considered as 20°C-32°C, the suitable period happens to be, meanly, *14 June* to *10 September* (i.e. 89 days) (Table 3). This period most early starts on *9 June* (Trabzon and Rize) and at latest *24 June*(Akçakoca) where ends most early on *29 August* (Akçakoca) and at latest on *18 September* (Sinop). On the other hand, this period lasts shortest in *Akçakoca* and longest in *Sinop* (Table 3). This period, in turn, happens between *13 May* and *17 October* [16], in Marmaris-Alanya coastal belt which includes the most important sea tourism centres of Turkey.

Precipitation

As selected stations, the mean annual rainfall in the Black Sea Region coastal belt ranges between 688,5 mm (Samsun) and 2244, 1 mm (Rize). The rainfall throughout the year falls mostly in autumn (35%) in which the frontal activities are strongest. This season followed by winter (29%) in which again the frontal activity is high. The air movements moving from northwest to southeast over Black Sea get higher in the northern slops of the Northern Anatolia Mountains and accordingly the summer again gets high rainfall (19%). The spring has the least rainfall percentage with 17% thanks to the dominance of the Southern originated winds, föhn winds than lessen the precipitation possibility and lessen thermal difference between the sea and land [17]. As may be seen, the entire the Black Sea Region coastal belt takes high rates of precipitation in all seasons and Black Sea precipitation characteristics are concrete. On the other hand, the least rainfall month is May, where the highest rainfall falls in October (Table 4).

Table 4. The mean monthly rainfall values and certain period rainfall rates of the meteorological stations selected from the Black Sea Region coastal belt of Turkey (1975-2008 period)

Meteorological stations	Mean monthly rainfall values (mm)												Rainfall rates in May-September period (%)	Rainfall rates in June-August period (%)
	I	II	III	IV	V	VI	VII	VIII	IX	X	XI	XII		
Akçakoca	112,3	74.4	74,3	56,3	48.6	58,7	74,3	89,3	89,9	140,2	123,7	131,3	33,61	20.71
Zonguldak	131.8	84.0	84,9	57,9	53,0	69,6	81,4	93,2	119,7	154,9	151,7	152,3	33,77	19,78
Amasra	102,9	67,9	66,1	48,2	42,6	64.4	61,2	77,6	101,5	117,4	119,4	119,0	35,14	20,56
İnebolu	111,1	84.6	72,0	45.4	52,7	47,6	45,3	59,1	92,6	141,8	132.5	140,1	29,01	14,83
Sinop	69,5	48,8	47.5	37.3	33,8	35.8	34,5	42,8	65,1	92,0	87.0	82,3	31,34	16,72
Samsun	60,7	50,7	56,8	58,6	49,3	48,6	30,2	37,5	52,3	87,6	83,0	73,2	31,64	16,89
Ordu	97,9	86,5	77,1	76,9	55,3	76,4	58,6	61,0	82,1	136,9	131.5	103,9	31,93	18,77
Giresun	118,0	94,9	90,4	85,1	69,0	84,3	69,0	89,0	120,9	172,2	150.7	121,1	34,17	19,16
Trabzon	76,8	64,8	59,0	59,6	53,0	54,5	37,1	51,5	74,1	119,3	95.7	85,9	32.50	17,21
Rize	212.4	180,6	141.6	92.0	99,0	135.9	141,2	180,0	243,6	309,9	257.6	250,3	35.63	20,37
Hopa	200,2	165,3	135.7	87.6	93,0	161.2	143,8	178,3	264,5	314,3	263,3	235,1	37,49	21,55
Mean	*117,6*	*91,1*	*82,3*	*64.1*	*59,0*	*76,1*	*70,6*	*87,2*	*118,8*	*162,4*	*145,1*	*135,9*	*33,3*	*18,8*

Source: Calculated and organized according to the data of the Ministry of Environment and Forestry, Turkish State Meteorological Service.

The rainfall rates in May-September period is 33, 3% and fall highest in Hopa (37, 49%) and least in ønebolu (29, 01%). When we look at the summer months, the average is 18, 8% with highest percentage in Hopa (21, 55%) and lowest fall in ønebolu (14, 83%); i.e. high rainfall rates even in summer (Table 4).

The region with highest number of precipitation days is again the Black Sea Region coastal belt in Turkey. The total number of days with precipitation is more than 130 and the mean is 151 days. Accordingly, the precipitation possibility is about 41%. That is to say there is the possibility of precipitation in each 2, 5 days. The precipitation possibility in mean is about from 50% (December) to 28% (July) in the selected stations. The least possibility of precipitation is seen in Middle Black Sea sub region coastal belt (36-38%), followed by Western Black Sea Sub region coastal belt (38-41%). The highest possibility of precipitation is in Eastern Black Sea coastal belt with 43-50%. On the other hand, the precipitation possibility changes so slightly through the year. The precipitation possibility is high even in summers. The most suitable station is Sinop in summer (July: 19% and mean summer: 23%). Rize is the station with highest precipitation rate (July: 47% and mean summer: 48%) and worth to attention. In mean seasonal precipitation possibilities are as follows [18]: Winter (49, 3%), spring (44, 7%), summer (31%) and autumn (41%).

Wind speed

According to the wind data of selected stations along the Black Sea Region coastal belt, mean yearly wind speeds vary between 1, 2 m/sec (Giresun) and 4, 5 m/sec (Amasra). It rests within threshold values presumed for sea tourism activities during May-September period. At any station it does not exceed 5, 4 m / sec (Table 5). The highest wind speeds were recorded in Amasra and ønebolu. The wind speeds are usually 1, 2 m/sec to 2, 2 m/sec, which indicates very low values in between Samsun and Hopa the coastal belt (Table 5).

Table 5. The mean monthly, annual and certain period wind speed values of the stations selected from the Black Sea Region coastal belt of Turkey (1975-2008 period)

Meteorological stations	Mean monthly wind speed values (m/sec)												Mean yearly wind speed values (m/sec)	Mean wind speed values in May-September period (m/sec)	Mean wind speed values in June-August period (m/sec)
	I	II	III	IV	V	VI	VII	VIII	IX	X	XI	XII			
Akçakoca	2,3	2,4	2,1	1,8	1,7	1,8	1,9	2,1	2,2	2,3	2,3	2,3	2,1	1,9	1,9
Zonguldak	2.8	2,8	2,6	2,4	2,0	2,0	2,1	2,2	2,4	2,4	2,6	2,8	2,4	2,1	2,1
Amasra	5,1	5,1	4,7	4,2	3,7	3,6	4,0	4,1	4,2	4,6	4,8	5,4	4,5	3,9	3,9
İnebolu	3,7	3,5	3,3	2,9	2,5	2,9	3,0	3,3	3,6	3,6	3,7	3,8	3,3	3,1	3,1
Sinop	3,5	3,7	3,5	3,2	2,7	2,8	3,1	2,8	2,8	3,0	3,0	3,3	3,1	2,8	2,9
Samsun	3,3	2,9	2,4	1,9	1,7	2,0	2,3	2,4	2,2	2,1	2,4	3,2	2,4	2,1	2,2
Ordu	1.7	1,7	1,6	1,5	1,5	1,6	1,8	1,9	1,7	1,6	1,6	1,6	1,7	1,7	1,8
Giresun	1,3	1,3	1,3	1,2	1,1	1,1	1,1	1,1	1,1	1,1	1,2	1,2	1,2	1,1	1,1
Trabzon	2,3	2,4	2,2	1,9	1,7	1,9	2,1	2,3	2,4	2,3	2,4	2,4	2,2	2,1	2,1
Rize	1.5	1,4	1,4	1,4	1,3	1,4	1,3	1,3	1,4	1,4	1,5	1,4	1,4	1,3	1,3
Hopa	3,8	3,2	2,5	2,0	1,5	1,7	1,7	1,8	2,1	2,2	3,0	3,9	2,5	1,8	1,7
Mean	*2,8*	*2,8*	*2,5*	*2,2*	*1,9*	*2.1*	*2,2*	*2,3*	*2,4*	*2,4*	*2,6*	*2,8*	*2,4*	*2,2*	*2,2*

Source: Calculated and organized according to the data of the Ministry of Environment and Forestry, Turkish State Meteorological Service.

Table 6. The mean monthly, yearly and certain period relative humidity values of the stations selected from the Black Sea Region coastal belt of Turkey (1975-2008 period)

Meteorological stations	Mean monthly relative humidity values (%)												Mean yearly relative humidity values (%)	Mean relative humidity values in May-September period (%)	Mean relative humidity values in June-August period (%)
	I	II	III	IV	V	VI	VII	VIII	IX	X	XI	XII			
Akçakoca	80,3	79,7	80,4	81,6	83,3	81,2	82,1	82,9	82,8	84,0	81,3	79,1	81,6	82,5	82,1
Zonguldak	67,3	66,8	66,9	67,5	70,9	69,1	71,4	72,2	70,8	71,9	66,9	66,4	69,0	70,9	70,9
Amasra	69,9	69,9	70,9	72,4	76,2	73,6	75,5	75,1	71,7	72,2	68,5	67,9	72,0	74,4	74,7
İnebolu	74,1	74,6	75,2	77,0	78,6	75,3	73,7	74,4	76,0	78,1	74,4	73,0	75,4	75,6	74,5
Sinop	70,5	70,5	73,3	75,5	77,4	75,2	75,5	75,1	74,5	74,1	71,5	69,8	73,6	75,5	75,3
Samsun	67,1	69,7	75,0	79,3	80,5	76,1	73,4	73,9	75,3	76,2	70,5	66,4	73,6	75,8	74,5
Ordu	67,7	69,3	72,8	75,3	76,1	72,4	72,6	72,7	73,3	74,8	71,0	68,0	72,2	73,4	72,6
Giresun	68,4	69,7	73,2	76,3	79,0	76,3	76,3	76,0	75,9	75,1	70,1	67,6	73,7	76,7	76,1
Trabzon	66,1	66,7	70,9	73,8	77,0	74,3	74,2	74,2	73,7	72,7	68,5	65,7	71,5	74,7	74,2
Rize	73,0	72,6	74,2	74,5	76,4	75,2	77,1	78,1	78,7	79,3	76,2	73,8	75,8	77,1	76,8
Hopa	61,7	65,3	70,0	73,2	76,9	76,4	78,5	78,9	77,0	74,7	66,6	60,3	71,6	77,5	77,9
Mean	*69,6*	*70,4*	*73,0*	*75,1*	*77,5*	*75,0*	*75,5*	*75,8*	*75,4*	*75,7*	*71,4*	*68,9*	*73,6*	*75,8*	*75,4*

Source: Calculated and organized according to the data of the Ministry of Environment and Forestry, Turkish State Meteorological Service.

The mean monthly wind speeds change between 1, 1 m/sec. (Giresun: May-October period) and 5, 4 m/sec (Amasra: December) in generally (Table 5). Between the September and March, through which the depression passes are frequent due to, in general, polar frontal activity, the wind speeds are higher however the mean monthly wind speeds demonstrate close values to mean annual. However, in the May -September period, in which sea tourism is intense, the mean monthly wind speeds fall under the mean annual and particularly in Eastern Black Sea sub region demonstrates between 1, 1 m/sec and 2, 4 m/sec speeds.

4. 5 Relative humidity

The meteorological stations selected in the Black Sea Region coastal belt demonstrate general mean monthly relative humidity values as varying from 68,9% (December) to 77,5% (May). The mean yearly relative humidity values in the coastal belt are 73, 6% and vary between 69, 0% (Zonguldak) and 81, 6% (Akçakoca). That is to say, the yearly relative humidity values are high (Table 6).

In the May-September period, in which the sea tourism activities are intense, the mean relative humidity values happen to be 75, 9% and climb to 75, 5% in summer season. The lowest values are close to the mean yearly in the mentioned periods. The lowest periodical mean relative humidity, among the selected stations, is 70, 39% (Zonguldak) for both the September-May and summer period. The highest mean relative humidity is 82, 5% and 82, 1% in Akçakoca respectively for September –May and summer periods (Table 6).

4.6 Sea surface temperature

When the ideal sea surface temperature as 22█- 25?C with respect to sea bathing is taken into consideration, the favourable period is seen in Rize (96 days) the longest, in ønebolu (34 days) the shortest period (Table 7). This period starts on 29 June (Akçakoca) the earliest, on 23 July (ønebolu) the latest, it ends on 26 August (ønebolu) the earliest, on 10 October (Rize) the latest (Table 7). On the other hand, this period on the study area is seen between *11 July* and *21 September* in mean.

Table 7. Ideal and favourable sea surface temperature periods for sea bathing of selected stations along the Black Sea Region coastal belt of Turkey (1975-2008 period)

Meteorological stations	Ideal sea surface temperature (22–25°C) periods Starting and termination date	Number of days	Favourable sea surface temperature (18–28°C)periods Starting and termination date	Numbe r of days
Akçakoca	29 June - 10 September	74	7 June- 21 October	137
Zonguldak	20 July - 7 September	50	18 June - 21 October	126
Amasra	4 July - 16 September	75	7 June - 17 October	133
İnebolu	23 July -26 August	34	14 June-16 October	125
Sinop	10 July - 14 September	67	14 June - 29 October	138
Samsun	6 July - 22 September	79	13 June - 28 October	138
Ordu	8 July-17 September	72	10 June-22 October	135
Giresun	18 July-30 September	75	22 June-4 November	136
Trabzon	11 July-9 October	91	10 June-7 November	151
Rize	7 July-10 October	96	14 June-11 November	151
Hopa	13 July-8 October	88	18 June-6 November	142
Mean	*11 July - 21 September*	*73*	*13 June - 28 October*	*138*

Source: Calculated and organized according to the data of the Ministry of Environment and Forestry, Turkish State Meteorological Service.

In case favourable sea surface temperatures are taken into consideration as 18°C-28°C, favourable periods with respect to sea bathing increase and vary from 125 days (ønebolu) to 151 days (Trabzon and Rize) (Table 7). Within the year, the favourable period is seen between *13 June* and *28 October* in the study area. This period starts on *7 June* (Akçakoca and Amasra; the earliest, *on 22 June* (Giresun) the latest, it ends on *16 October* (ønebolu) the earliest, on *7 November* (Trabzon) the latest (Table 7).

4.6 Discussion

By means of sunshine duration which is important for the sea tourism, the most suitable period is the two months between mid June and mid August in the Black Sea Region coastal belt. However, the most suitable area for this parameter is the coastal belt between Akçakoca and Samsun. The coastal belt between Samsun and Hopa does not demonstrate a certain period for sunbathing opportunities.

The sunshines reach to the study area with a 50º-71º angle in May-September period. However, the weather being generally misty, rainy and cloudy falls the sunshine durations in the period of high sea tourism season. Thus, the shine durations lessen. Moreover, the sunbathing comfort decreases in summer months in which the precipitation possibility and cloudiness rates are high.

When we consider that the tourists of the OECD countries have stated that their mean monthly temperature of holiday preference is 21ºC, it can be seen that all the stations in the study area are suitable for such figure. Particularly the months July and August have the most suitable conditions.

The mean wind speeds in the Black Sea Region coastal belt do not exceed 6 m/sec throughout the year. Thus, it can be said that all the selected stations contain suitable conditions. The relative humidity values are high all over the year and more than 70%. The relative humidity values are especially high in September-May and summer periods in which sea tourism activities are intense. The lowest relative humidity values have been spotted in Zonguldak.

The most important problematic climate factor for the sea and sunbathing is precipitation. The rainfall and precipitation possibility is quite high even in summer months. The most suitable coastal belt by means of precipitation is again Akçakoca-Samsun coastal belt. The high summer rains between Samsun and Hopa generate cause significant problems.

If the ideal air temperature threshold values for sea bathing are regarded as 25°C -28°C and ideal sea surface temperature threshold values are regarded as 22°C -25°C, then, the Black Sea Region coastal belt air temperatures do not exceed daily 25°C in mean and accordingly no certain ideal period may be shown. However, if the sea surface temperature threshold values as 22°C - 25°C together with air temperature threshold values as 20°C -32°C, then 62 days between *11 July* and *10 September* may be said to be suitable for sea bathing.

If the suitable air temperature threshold values for sea bathing are regarded as 20°C-32°C and suitable sea water temperature threshold values are regarded as 18°C -28°C, then 86 days between the *17 June* and *10 September* may be said to be suitable for sea bathing (Table 8). This period starts earliest on *10 June* (Trabzon) and latest on *24 June* (Akçakoca) in the Black Sea Region coastal belt. It ends earliest on *29 August* (Akçakoca) and latest on *18 September* (Sinop). On the other hand, this period lasts shortest in Akçakoca (*67 days*) and longest in Trabzon (*99 days*) (Table 8). This period, in turn, in Marmaris-Alanya coastal belt happens between *13 May* and *17 October* [16], in Alanya-Samanda÷ coastal belt between *9 May* and *23 October*.

Table 8: Favourable periods for sea bathing of selected stations along the Black Sea Region coastal belt of Turkey (1975-2008 period)

Meteorological stations	Favourable periods for sea bathing Starting and termination date	Number of days
Akçakoca	24 June - 29 August	67
Zonguldak	18 June - 2 September	77
Amasra	13 June -30 August	79
İnebolu	23 June - 2 September	72
Sinop	14 June - 18 September	97
Samsun	13 June - 16 September	96
Ordu	13 June- 16 September	96
Giresun	22 June- 16 September	87
Trabzon	10 June- 16 September	99
Rize	14 June- 16 September	95
Hopa	18 June- 8 September	83
Mean	*17 June-10 September*	*86*

Source: Calculated and organized according to the data of the Ministry of Environment and Forestry, Turkish State Meteorological Service.

CONCLUSION

Despite the problems arising from short sunshine durations, high rainfall, precipitation possibility and relative humidity, 86 days (between *17 June* and *10 September)*, and particularly the month July) in the Black Sea Region coastal belt are suitable for sea and sun bathing throughout the year. The most suitable coastal belt by this means is between Akçakoca and Samsun. The coastal belt between Ordu and Hopa is not so suitable for sea and sun bathing.

The mean temperatures do not exceed 25°C in general and the highest mean annual temperatures vary between 26,4°C (Akçakoca) and 28,2°C (Ordu) and do not go over 30°C. Thus, it is an important alternative to Aegean and Mediterranean coasts where high temperatures exist especially in summer season. The Black Sea Region coastal belt is close to the important centres, Istanbul and Ankara in the first place and thus has an important advantage. However, high relative humidity and summer rains may generate some problem for sea tourism.

The analyses in this study have been performed by basing on the daily and monthly data of the stations. Performing this study for all the beaches in the Black Sea Region coastal belt and examining the day hours suitable/not suitable for sea and sun bathing, should be considered important for tourism planning future of the coastal belt. Therefore, establishing stations all over the coastal belt beaches capable of performing reliable and permanent observations, improving the works of the present stations and communicating the obtained data to the tourists via mass media and meteorological information plates are required. Such coastal stations to be established shall be capable of observing and assessing air temperature, sea surface temperature and wind speed for sea bathing purposes; sunshine duration, solar severity, cloudiness, temperature, humidity, wind speed, thermal comfort and UV radiation for sunshine purposes; and sand surface temperature, wind speed and albedo for sand purposes [19].

Tourism activities should be supported by alternative tourism events for to minimize the problems arising from the climate. This is possible since the Black Sea Region coastal belt has many attractions and suitable locations many activities such as trekking, sightseeing and photo safari etc. many activities. Thus, the infrastructure and superstructure problems, transport in the first place, should be solved immediately

REFERENCES

1. Wall G. The tourism industry and its adaptability and vulnerability to climate change. In: Amelung B, Blazejczyk K, Matzarakis A, editors. *Climate change and tourism – assessment and coping strategies*, Maastricht-Warsaw-Freiburg; 2007: 5-19.
2. Özgüç N. *Tourism Geography. Features. Regions.* Istanbul: Çantay Bookstore, 1998. (In Turkish)
3. Zaninoviü K, Matzarakis A. Biometeorological basis for tourism. In: Matzarakis A, Freitas CR, Scott D, editors. *Developments in Tourism Climatology*; 2007: 24-28, http:// www. Klimazwei .de/ Portals /0/dtc_ 2007_ zaninovicmatzarakis. pdf, Last access date: 15.05.2009.
4. de Freitas C. R. Theory, concepts and methods in tourism climate research. In: Matzarakis A, Freitas CR, editors. *Proceedings of the first international workshop on climate, tourism and recreation, Report of a workshop,* Porto Carras, Neos Marmaras, Halkidiki, Greece, 2001: 3-21.
5. Do÷aner S. *Tourism geography of Turkey.* Istanbul: Çantay Bookstore, 2001. (in Turkish)
6. Atalay ø, Mortan K. *The regional geography of Turkey with map and illustrations.* Extended 3rd ed. Istanbul: ønkılâp Bookstore, 2006. (In Turkish)
7. Atalay ø. *Introduction to geomorphology of Turkey*. Izmir: Ege University Faculty of Arts Publications; 9: 1987. (In Turkish)
8. Amelung B, Viner D. The vulnerability to climate change of the Mediterranean as a tourist destination. In: Amelung B, Blazejczyk K, Matzarakis A, editors. *Climate change and tourism – assessment and coping strategies*, Maastricht-Warsaw-Freiburg; 2007: 41-55.
9. Corobov R. Climate change and Moldova's tourism: Some indirect consequences. In: Amelung B, Blazejczyk K, Matzarakis A, editors. *Climate change and tourism – assessment and coping strategies*, Maastricht-Warsaw-Freiburg; 2007: 173-189.

10. Ülker ø. *Health Tourism. Resource. Planning. Presentation.* Ankara: Republic of Turkey, the Ministry of Tourism Puplishing; 1994. (in Turkish)
11. Güneú A. Sweltering weather and meteorological elements creating sweltering weather, determination of sweltering weather. In: *Book of Abstracts on Medical Biometeorology Seminar*, Ankara: Turkish State Meteorological Service Publications; 1983: 17–18. (In Turkish)
12. Mieczkowski Z. The tourism climatic index: A method of evaluating world climates for tourist. *The Canadian Geographer/Le Géographe Canadien*, 1985; 29(3): 220–233.
13. Koçman A. *Aegean plains climate in terms of human activities and environmental effects.* Izmir: Ege University Faculty of Arts Publications; 73: 1993. (In Turkish)
14. Dönmez Y. *General climatology and climate studies.* østanbul: Istanbul University, Institute of Geography Publications; 102; 1984. (In Turkish)
15. Koçman A. *Turkey Climate*. Izmir: Ege University, Faculty of Arts Publications; 72; 1993. (In Turkish)
16. Güçlü Y. The determination of sea tourism season according to the climatic conditions in Marmaris-Alanya coastal belt (SW of Turkey). *Journal of tourism challenges and trends,* 2008; 2: 91-113.
17. Erinç S. *Climatology and methods*. Istanbul: Istanbul University, Institute of Marine Sciences and Geography Publications; 2; 1984. (In Turkish)
18. Güçlü Y. *Climate comfort for coastal tourism in the sea coastal belts of Turkey*. 1st ed. Istanbul: Lisans Publishing; 2010. (In Turkish)
19. Matzarakis A. Weather and climate-related information for tourism. *Tourism and Hospitality Planning and Development,* 2006; 2: 99–115.

Chapter 9

Natural and Traditional Defense Mechanisms to Reduce Climate Risks in Coastal Zones of Bangladesh

Mohammed Ataur Rahman[a,*], **Sowmen Rahman**[b]

[a] *Centre for Global Environmental Culture (CGEC), IUBAT—International University of Business Agriculture and Technology, 4 Embankment Drive Road, Uttara Model Town, Sector 10, Dhaka 1230, Bangladesh*

[b] *Department of Urban and Regional Planning, Jahangirnagar University, Savar, Dhaka, Bangladesh*

ABSTRACT

Substantially resourceful and densely populated coastal zones of Bangladesh experience numerous extreme events linked to hydro-meteorological processes viz. cyclones, tidal surges, floods, salinity intrusion and erosion etc. These hazards give rise to extensive damage to property and loss of lives every year. Further, anthropogenic activities in the coastal zones are accentuating environmental degradation causing widespread suffering. Cyclones and tornadoes in particular damage infrastructures and crops every year affecting the economy of the country negatively. Some naturally adapted plants as well as landscapes usually reduce the speed of cyclones and tornadoes and thus, protect the coastal zones. However, human activities have destroyed many of the forests and landscapes. Sundarbans and Chokoria Sundarbans mangrove forests of Bangladesh are under a great threat of extinction due to illicit logging and agricultural expansion. At least 34 plant species of tropical forest are on the verge of extinction.

Many animals e.g., cats, bears, porcupines, wild boars, pythons and anteaters are in the process of being wiped out from the coastal areas. Among the marine and coastal species, Red crabs, jelly-fish, sharks, and dolphins are also rare but these were the major species prior to 1980s. This study revealed that during the recent decades there has been massive plantations and construction of embankment and polderization but these and other measures have been found to be impractical and ineffective in reducing disasters in coastal areas. There is a need for integration of traditional coping practices and wisdoms with modern approaches. Available knowledge on some of these traditional practices has been documented for establishing a sustainable policy for management of coastal zones of Bangladesh. By combining traditional and scientific management of coastal ecosystem with mangroves and other plants following triple-tier mechanism and habitat, it is possible to reduce the effects of natural and climate change-induced disasters. Under such a management system, the entire coastal zone can be made more productive and sustainable.

Keywords: Coastal zone; Mangroves; Excavation of fresh water reservoir; Triple-tier mechanism

1. INTRODUCTION

In South Asia, Bangladesh is the most densely populated delta of the Ganges–Brahmaputra–Meghna (GBM) basin. Most of the country is the flood-plain of the GBM river systems with their tributaries and distributaries. The fertile alluvial flood-plain is characterized by gently undulated landscape with hills and hillocks in the north, east; a central undulated red soil terrace, the Madhupur Tract; and a huge coastal zone with highly fertile land, rivers, estuaries, mangroves, seashore and islands adjacent to the land-water interface of the southern part of the country. The coastal zone comprises 19 administrative districts that have a great diversity of natural resources including coastal fisheries, forests, salt, and minerals, as well as high potential for exploration of both onshore and offshore natural gas. It harbors ports, tourism facilities, and other development opportunities (Michel and Pandya, 2010). Although Bangladesh is most vulnerable to frequent hydro-meteorological hazards such as floods, cyclones and droughts, historically people had developed coping mechanisms including the use of traditional practices for making their homes and homesteads resilient to floods, tornadoes and erosion etc.; climatic-season-based cropping, fish-farming and developed major transportation by boat. They applied traditional knowledge for agriculture, floodplain management and used natural and traditional defense mechanism etc. However, over the last few decades, along with the increased frequencies of climate change disasters, increased urbanization, unplanned road construction, industrialization and population growth, many aspects of traditional life have been changing very fast (Nagaraju, 2012 and Raygorodetsky, 2013).

Coastal zone communities in different parts of the world, have long before conventional western practices were introduced, relied on their traditional knowledge to reduce disaster risks. Recent human history contains examples of entire islands rendered uninhabitable through environmental destruction owing to external causes; small island developing states (SIDs) are fully aware about the environmental consequences of ill-conceived development with catastrophic effects. Unsustainable development threatens the livelihood and the cultures they nurture; deforestation, coral reef deterioration, habitat degradation and loss, and the introduction of certain non-indigenous species are the most significant causes of the loss of biodiversity and endemism in SIDs which cause fragility of island ecosystems. However, an improved capacity to tap local knowledge and to develop environmentally sound endogenous technologies is also an important step towards sustainable development in a number of areas, including agriculture, agricultural processing, construction, communications and the marine sciences (UNEP, 1994, GDRC, 2014 and FAO, 2014).

Traditional customs and management structures are still strong in many coastal communities throughout Timor-Leste. Traditional natural resource management schemes have been weakened by centuries of suppression by Portuguese colonization and Indonesian occupation. Development, modernization, and education have been influencing coastal communities and further weakened traditional coastal resource management; the younger generation is losing such valuable local and traditional knowledge. There is an urgent need to collect and record information on traditions and customs to avoid loss of indigenous culture and heritage that can also serve as model for sustainable use of natural resources (Wever, 2008).

However, Bangladesh is recognized globally as most vulnerable to climate change extremes (Climate Change Cell, 2007). About 80% of the country is deltaic floodplain of the GBM with many rivers flowing from the central India, Himalayas, China, Assam, Lusai and Arakan-Burmese ranges and the catchments of the GBM Rivers flowing to the Bay of Bengal through the estuaries (Fig. 1). Bangladesh is the most vulnerable country in the world to tropical cyclones and the sixth most vulnerable to floods (Government of Bangladesh, 2008). More than 68 million people have been directly affected over the last eight years (Vos et al., 2010), and millions of lives and livelihoods are threatened by frequent weather-related disasters. Low-lying lands, coastline areas and floodplains of most part of the country are highly exposed to both disasters and sea level rise, especially in the coastal zones (Global Humanitarian Forum, 2009).

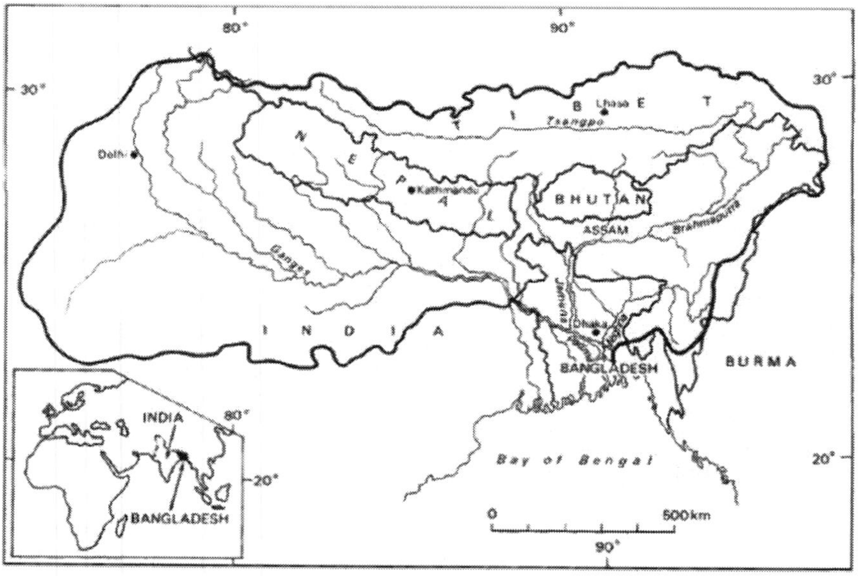

Figure 1. Catchments of the Ganges–Brahmaputra–Meghna Rivers flowing to the Bay of Bengal. *Source:* Brammer, 1990.

Above the coastal zone, one-third of the country is partially elevated plain-land, which gets flooded temporarily but with the increased population this floodplain has gradually been occupied with expanded homes and townships. Another 44,000 km^2 area consists of wet-bodies which remain wet for most of the time. However, after the 1960s Green Revolution, these wetlands were also destroyed by earth-filling for agricultural expansion especially for High Yielding Variety (HYV) rice developed by International Rice Research Institute (IRRI). Thus the natural systems including forests, wet-bodies and traditionally managed floodplains and coastal ecosystem have been destroyed and the country is facing serious climate change disasters affecting millions every year (Flood Archive, 2004).

Considering the increasing frequencies of climate change disasters, this study was conducted to compile the natural and traditional adaptation practices and the defense mechanisms used by the people of the coastal zone of Bangladesh to reduce vulnerability to hydro-meteorological hazards and to find out sustainable coping methods which can be integrated in the national or regional policy.

During the study, hazards and degradation in coastal zone, traditional disaster risk reduction measures, mal-adaptation practices and disasters and purview of policy challenges have been assessed.

2. METHODOLOGY

During the study, relevant information was collected from grey and published literatures of different scientific research work. Information was also gathered through organizing local workshops and seminars; attending different regional workshops and conferences; visiting research institutions and meteorological stations. Random interviews of different stakeholders, rural and urban administrative bodies, Non-Government Organizations (NGOs), men and women and youth communities of the coastal region, were conducted. Media sources, folklore and local cultures were another source of information.

The effects of climate variability and change, and vulnerability of natural and human systems in the coastal area were studied. For a cross-scale synthesis and policy recommendation, the traditional practices and coping behaviors were investigated; their efficacies were highlighted and compared with present practices to find out the misfits as well as to integrate the scientific basis of the traditional knowledge regarding natural and artificial defense mechanism followed by the people in this most vulnerable region of the earth for thousands of years.

3. HAZARDS AND DEGRADATION IN COASTAL ZONE

The coastal zone has many resources like fertile land, fishing, mangrove forests, marine and terrestrial biodiversity, scenic beauty, marine resources, meeting point of flora and fauna of fresh and saline water;

salt fields, and minerals: Quartzes and Zircon, Uranium etc.; and easy transportation and sailing facilities (Rahman, 2010 and DoE, 2013). These have facilitated developments of ports, industries and tourism.

The GBM estuaries in the south; the Karnaphuli, Halda and Sangu rivers and the shoreline of the Arakan ranges in the southeast form a distinct feature of the whole coastal zone of Bangladesh (Fig. 1). It has a difficult coastline with many rivers and distributaries and complex ecology which is affected by natural hazards like cyclone, coastal flooding, tidal surges, erosion, salinity and human related activities. About 50 million people, nearly one-third of the total population of Bangladesh live in the coastal zone (Miyan, 2009 and Rahman, 2010). However, some phenomena often create disasters; endanger and disrupt lives and the whole coastal ecosystem. Among these, tropical cyclones and tornadoes, tidal surges and floods, erosion, heavy siltation, and pollution especially from the mega-cities and ports, shrimp hatchery and shrimp farms are the most prominent. Deforestation, over-fishing and cutting of hills for unplanned construction, ship-breaking industries and tourism have accelerated destruction of the ecosystem. Unregulated removal of sand, gravel and pebble deposits from beaches and underwater coastal slopes add to the ongoing threats to the coasts (Banica et al., 2003, Bird, 1979 and Rahman, 2011a). More than 34 species of tropical rainforest plants, including *Podocarpus nerifolia* and *Enteda phaseoloids* are facing extinction from the coastal hill forests of Chittagong ranges (Rahman, 2011a and Rahman, 2011b; Khan et al., 2001, Hossain and Ahmed, 2008 and Miyan, 2012). Many animals such as cats, bear, porcupine, wild boars, pythons and anteater are in the process of being wiped out from the coastal areas. Among the marine and coastal species, Red crabs, jelly-fish, sharks, and dolphins are becoming rare but these were the major species before 1980 (Rahman, 2011a). Coastal resource depletion leads to frequent conflict between users and causes serious socio-economic and cultural problems, such as weakening of the social fabric, marginalization, unemployment and destruction of property by erosion.

3.1. Recent trend of environmental changes and decline of resources

Human pressure, misuse of resources, introduction of inappropriate high input-based agricultural technologies; industrial and agricultural expansion and pollution are the characteristics of recent changes in the coastal zone (Sherbinin et al., 2007). Vulnerabilities in the coastal zone are increasing with accentuations of natural hazards caused by environmental degradation, climate change and human activities including exploitation of mangroves (Rahman, 2010 and Scialabba, 1998). Bangladesh has already been affected by land erosion, salinity intrusion (Seal and Baten, 2012) and loss of biodiversity. The potential threats are going to be even stronger in future (Pender, 2008 and Wong et al., 2007). Due to climate change effects, the incidences of tropical storms and tidal surges have increased in the coastal belts of Bangladesh, India, Myanmar and Sri Lanka. Since the frequency of cyclones in the Bay of Bengal is about 5 to 6 times the frequency in the Arabian Sea (IMD, 1979), the Bay of Bengal share comes out to be about 5.5%. The breakdown of this 5.5% for the littoral countries of the Bay of Bengal is given in the last row of Table 1. Bangladesh is hit by about 0.93% (~1%) of the world's total tropical storms, India by 3.34%, Myanmar by 0.51%, Sri Lanka by 0.22%, and 0.50% die in the Bay without hitting any country. These numbers were arrived at by considering the 110. However, 9.5% depressions hit Bangladesh (Table 1).

Table 1. Number of cyclones forming in the Bay of Bengal and hitting the littoral countries (1877 to 1995).

Type	Bangladesh	India	Myanmar	Sri Lanka	Dead	Total
All types	154	848	71	35	115	1223
Depressions	68	539	24	15	69	715
CS (cyclonic storm)	43	197	23	12	35	310
SCS (severe cyclonic storm)	33	112	24	8	11	198
CS+SCS	86	309	47	20	46	508
Percent of global total CS+SCS	0.93	3.34	0.51	0.22	0.50	5.5
Percent depressions	9.5	75.38	3.36	2.1		

If the world's tropical cyclones with death tolls in excess of 5000 are considered, it is found that 16 out of the 35 such disasters occurred in Bangladesh and 11 in India (Table 1). About 53% of the world deaths from these cyclones took place in Bangladesh and about 23% in India, for a combined total of 76% in these two countries. Bangladesh and India suffer most, although both of them together are hit by only 4.27% of the world storms, Serious and recurring floods have taken place during 2002, 2003, and 2004. Cyclones originating from the Bay of Bengal have been noted to decrease since 1970 but the intensity has increased (Ali, 1999).

The cyclones such as Sidr-2007, Nargis-2008 and Aila-2009 are the typical examples which have caused significant damages to lives and properties of the coastal zones (Table 2) (Dube, 2009).

Table 2. Tropical cyclones affected coastal zones

Cyclones	Affected Regions	Wind Speed (km/h)	Death	Damages (million US $) in Bangladesh
Bhola cyclone-1970	Bangladesh, India	205	500,000 (in Bangladesh)	86.4
Bangladesh cyclone-1991	Bangladesh	260	138,000	1,500
Sidr-2007	Bangladesh	260	4,036	1,700
Nargis-2008	Myanmar, India, SriLanka, Bangladesh	215	138,366 (126 in Bangladesh)	10,000
Aila-2009	Bangladesh, India	120	325 (26 in Bangladesh)	552.6
Mahasen-2013	Bangladesh	88	17 (in Bangladesh)	200

Withdrawal and diversion of upstream water from the major river systems in the dry monsoon have drastically reduced the freshwater flow. A decrease of about 82% is found in case of the minimum discharges during the dry months in the Ganges only (Afroz and Rahman, 2013). As a result, the salinity has been increased; salinity intrusion affects the coastal agriculture, natural fish-breeding centers and fresh water supply to the urban and rural areas of the coast. More than two hundred international rivers are carrying waters into Bangladesh from the neighboring countries. Withdrawal of water through upstream dams and embankments has decreased the dry season flow resulting in salinity intrusion which in-turn hampers agriculture in the coastal zone (Bangladesh State of the Environment, 2001). A regional consensus is needed for the share of common water flow.

Coastal populations, including three million inhabitants of the 72 offshore islands are extremely vulnerable. About 18% of households of the southwestern coastal zone are dependent on resources of the Sundarbans, viz. shrimp-fry, honey, Golpata, mollusks, shell, crabs and medicinal plants etc. Because of increasing rough weather in the Bay, nearly 0.5 million people dependent primarily on fishing are in danger by losing their works. Over 160,000 coastal fishermen and 185,000 shrimp-fry collectors involved in marine fisheries are threatened due to cyclonic disasters and tidal surges and reduced fish and aquatic resources (NPDM, 2015). According to a recent study of Soil Research Development Institute (SRDI), Sundarbans has lost 8.3% land (about 50,000 ha) on its northern front due to deforestation for shrimp culture during the period of 2000 to 2010. This destruction has been done by clear-felling of mangrove trees and creating pockets of un-forested areas (The Daily Star, 2013). As a result, these pockets have become low depression zones and more vulnerable to climate-induced cyclones and tidal surges.

3.1.1. Observed effects of global warming

According to IPCC, 2007 (AR4), average temperature has registered an increasing trend of about 1 °C in May and 0.5 °C in November during the 14 years period from 1985 to 1998 in Bangladesh. It was also reported that decadal rainfall anomalies are above long-term averages since 1960s (Khan, 2000 and Mirza, 2002).

For the Indian monsoon, extreme rain events have an increasing trend between 1901 and 2005 and the trend is much stronger after 1950 (Rajeevan et al., 2008) and widespread increases in heavy precipitation in high elevated Himalayas foothills extending south into the Ganges basin particularly during summer (Roy, 2009) (Fig. 2 and Fig. 3).

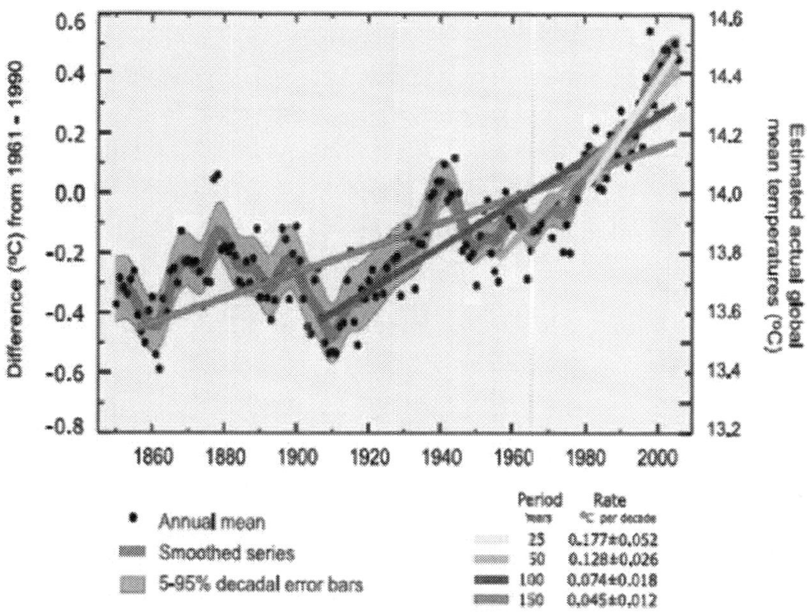

Figure 2. Rise in temp over 100 years 0.74 °C. Last 50 years decadal rate of rise 0.026 °C but doubles to 0.052 °C in last 25 years (*Source*: DLS, 2013).

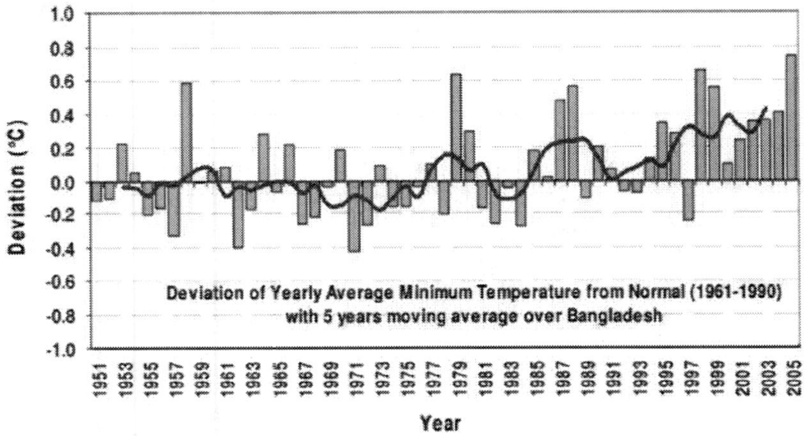

Figure 3. Deviation of yearly minimum temperature from normal (*Source*: DLS, 2013).

The annual mean rainfall exhibits increasing trends in Bangladesh (Table 3 and Table 4). Decadal rain anomalies are above long term averages since 1960s. Severe and recurring floods have taken place during 2002, 2003, and 2004. The intensity of cyclones originating from the Bay of Bengal has increased since 1970 which caused immense sufferings to lives and destruction of structures and natural resources (Cruz et al., 2007 and Sivakumar and Stefanski, 2011).

Table 3. General circulation models (GCM). Source: Climate ChangeCell,2009.

Year	Average temperature			Temperature increase			Average precipitation			Precipitation increase		
	Week °C	Month	Ave.	Week °C	Month	Ave.	Week Mm/month	Month	Ave.	Week Mm/month	Month	Ave.
1990	18.9	28.7	25.7	0.0	0.0	0.0	12	418	179	0	0	0
1930	21.4	29.4	27.0	1.3	0.7	1.3	18	465	189	6	47	10
2075	22.0	30.4	28.3	2.1	1.7	2.6	00	530	207	−12	112	28

Table 4. Climate change scenarios for Bangladesh. Source: Climate ChangeCell,2009.

Year	Temperature change °C mean		Precipitation change % mean		Sea level rise (cm)		
	Monsoon season	Dry monsoon	Monsoon season	Dry monsoon	SMC	NAPA scenario	IPCC 3rd upper range
2030	0.8	1.1	+6.0	−2.0	18	14	14
2050	1.1	1.6	+8.0	−5.0	30	32	32
2100	1.9	2.7	+12.0	−10	60	88	88

With the changing climate and withdrawal of water in the upstream, an acute water shortage has impacted badly the rapid urbanization, intense agriculture and population growth. Salt water from the Bay of Bengal is reported to have penetrated more than 100 km in the inland along tributary channels during the dry season (Sultana et al., 2008). The precipitation declines and droughts have resulted in drying up of wetlands and severe degradation of ecosystem services (IPCC, 2007). Sea level rise will worsen coastal erosion. In some coastal areas, a 30-cm rise in sea level can result in 45 m of landward retreat (Rabbani et al., 2010a and Rabbani et al., 2010b). Coastal erosion, cyclones, and storm surges will place coastal infrastructure—housing, industrial facilities, energy and sanitation systems, transportation and communication networks, tourist and cultural sites increasingly at risk (World Bank Report, 2013). In the southern region of Bangladesh it is projected that a 65 cm sea level rise by 2080s, will result in loss of 40% of the productive land. About 20 million people of the coastal area have already been affected by salinity. Rising sea levels and more intensive cyclones and storm surges could intensify the contamination of ground water and surface water causing more diarrhea outbreaks (Rabbani et al., 2010a and Rabbani et al., 2010b).

3.1.2. Cyclones and tidal surges

Owing to climate change effects the increased climate extremes of tropical cyclones along with tidal surges have severely affected the coastal zone (Table 2).

3.1.3. Floods

About 45.5 million people are exposed to severe and moderate floods like river flood, flash flood and tidal flood. Floods of 1974, 1987, 1988 and 1998 caused death of 30,000, 1657, 2379 and 1000 lives respectively and damaged crops and infrastructures (Climate Change Cell, 2009). Recurring floods during 2002, 2003, and 2004 caused huge damages of crops, structures and road transportation (Flood Archive, 2004 and Hossain, 2002). In 2007, flood inundated 32,000 km^2 area twice in July–August and in September and 16 million people were affected in three million households of which 85 thousand houses were totally damaged and more than one million damaged partially.

3.2. Effects of environmental change: Natural and anthropogenic

With the environmental change, agriculture has been suffering from uncertainty, for instance, cropping has been hampered by erratic precipitation patterns mainly of rainfall and effects of floods, droughts, salinity, cyclones and hailstorms. It has been predicted that rice and wheat production will drop by 8% and 32% respectively by 2050 (Bhattacharya et al., 2013) and fisheries are also expected to be impacted negatively.

Increased precipitation brings more water in the catchments which are beyond the drainage capacity, causing damage of infrastructure and drainage congestion in the urban areas due to faulty drainage and insufficient channels (Biswas and Emmelin, 2010). Salinity intrusion hampers irrigation and supply of portable water. Trans-boundary withdrawal of water disrupts hydrological cycle and cause increased rainfall during the pre-monsoon, wet monsoon and post-monsoon in upper catchments and/or within Bangladesh leads to more floods and water-logging; causes more river bank and coastal erosion (Climate Change Cell, 2009, MoEF, 2001, Rahman, 2004a and Rahman, 2004b).

However, over the years, considerable investment has been made in disaster management especially for flood control works using government and donors' fund and expertise. These measures have blocked traditional navigation routes and roads and highways have further worsened the situation. Embankments and polderization and unplanned afforestation have impacted negatively the coastal zones. Poor maintenance and inadequate management of the polders have contributed to internal drainage congestion and heavy external siltation. As a result, soil fertility and agriculture production in some areas are declining because of water-logging and increased salinity inside the polders (BWDB, 2013). In 2008, the Institute of Water Modeling (IWM) studied the impact of sea level rise in coastal river of Bangladesh and assessed the change in the tidal characteristics of the surrounding rivers due to sea level rise and its impact on inundation area of the polder. Studies revealed that high water level at the surrounding rivers of polders increases in the range of 30–80 cm for sea level rise of 32 cm and 88 cm respectively and this hampered the smooth drainage of the polders.

Prolonged water-logging affected cropping, thick wind-barriers increase the wind-speed and damaged the structures; uprooted trees damaged buildings, transmission lines, bridges and agricultural crops. However, increased awareness has significantly reduced the death tolls (Rahman, 2011a and Global Humanitarian Forum, 2009).

3.3. Future projections

PRECIS (Providing REgional Climates for Impacts Studies) generated scenarios for Bangladesh and predicted that pre-monsoon rainfall will be reduced but wet monsoon and post-monsoon rainfall will be increased, from 2051 and onwards. Monsoon rainfall will follow higher increasing trend and annual average rainfall will follow an increasing trend. Regarding temperature, monthly average maximum temperature will increase in monsoon period and will decrease in other periods but minimum temperature will increase in all periods. However, annual maximum and minimum temperature will follow an increasing trend (Climate Change Cell, 2009).

Impact on monsoon flooding will be severe and flood duration will be prolonged by significant number of days and inundation area and inundation depth will be increased (Climate Change Cell, 2009). The potential land loss estimated by IPCC is even worse—29,846 km^2 of land will be lost, making 14.8 million people landless by a sea level rise of 1 m. Land loss leads to loss of agricultural land, loss of homestead, loss of road and other communication infrastructure and above all loss of wide range of biodiversity. The great mangrove forest Sundarbans, the world heritage will be lost and the coastal region as well as the internal territory will be further impacted by the tropical cyclones (Earth Justice ACJP, 2009, IPCC, 2001 and OECD Report, 2003).

4. TRADITIONAL DISASTER RISK REDUCTION MEASURES

Since the Tropic of Cancer passes through the center of Bangladesh, the country is very prone to tornadoes due to wide-spread terrestrial wet-bodies all over the country. But the coastal zone is more prone to tornadoes and cyclonic storms (Bangladesh State of the Environment, 2001). The coastal zones have excellent natural defense systems with Arakan hill ranges along Chittagong and Cox's Bazar shoreline, strong windbreaks with hill forests in the eastern coast of the Bay of Bengal. Sea currents, web and tides and upstream fresh water flows play very important roles in enriching the coastal zones of Bangladesh (Rahman,

2010). The Sundarbans is the world's largest Mangrove forests. The National Park of Bangladesh was declared a Tiger Reserve in 1973, UNESCO World Heritage Site. Sundarban Biosphere Reserve was formed in 1989 through a notification for coordinating and integrating diverse activities of conservation, research and training for creating a better situation of harmony between man and environment as a part of Men and Biosphere Programme (MAB) (Sundarban Biosphere Reserve, 2006). Sundarbans' green cover has been standing firm against natural disasters for ages (Chand et al., 2012 and The Sundarbans, 1997). The vegetation consists of 64 plant species, that have the capacity to withstand estuarine conditions and saline inundation on account of tidal effects and can face strong winds. Sundri (*Heritiera fomes*), Gewa (*Excoecaria agallocha*), Kankra (*Bruguiera gymnorrhiza*), Passur (*Xylocarpus mekongensis, Rhizophora* spp., Sundri (*H. fomes*), Goran (*Ceriops decandra*) and Baen (*Avicennia alba*) are some of the important species. The Sundarbans mangrove forests with triple tier natural protection mechanisms protect the inlands from cyclonic storms originating in the Bay of Bengal (Rahman, 2010 and Miyan, 2012). Most of the plant species are adapted to the environment and can protect themselves as well as prevent the mounting wind pressure of the cyclones, and storm surges. Viviparous germination, pneumatophore, stilt roots, buttresses, floating behavior of drupes and pods are some of the unique natural adaptation features of many mangrove species Sanjay and Balaji, 1994. Some trees such as Baen and Sundri are so strong that they can easily resist 11,100 psi load (Sattar and Bhattacharjee, 1987). Tornado-prone zone areas (Fig. 4) have specially adapted growing bamboo clumps which are extremely strong with high elasticity and can easily disband the tornadoes. Therefore, every homestead has bamboo clumps in the west and or north-west sides (Rahman, 2004a and Rahman, 2004b). Bamboos also protect erosion with their strong anchorage-fibrous root system (Geotech Guideline, 2005). Similarly many palms and nuts also break the wind speed and protect the habitats and the lives from disasters especially, the cyclones and erosion (Bird, 1979). Local communities developed many adaptation techniques, innovations and knowledge to address climatic vulnerability. Rural people, farmers and others developed their location specific knowledge and practices of agriculture, natural resource management, human and animal health care. They know far more about their environment, livelihoods and how their society functions than those from the outside (BARCIK Report, 2013).

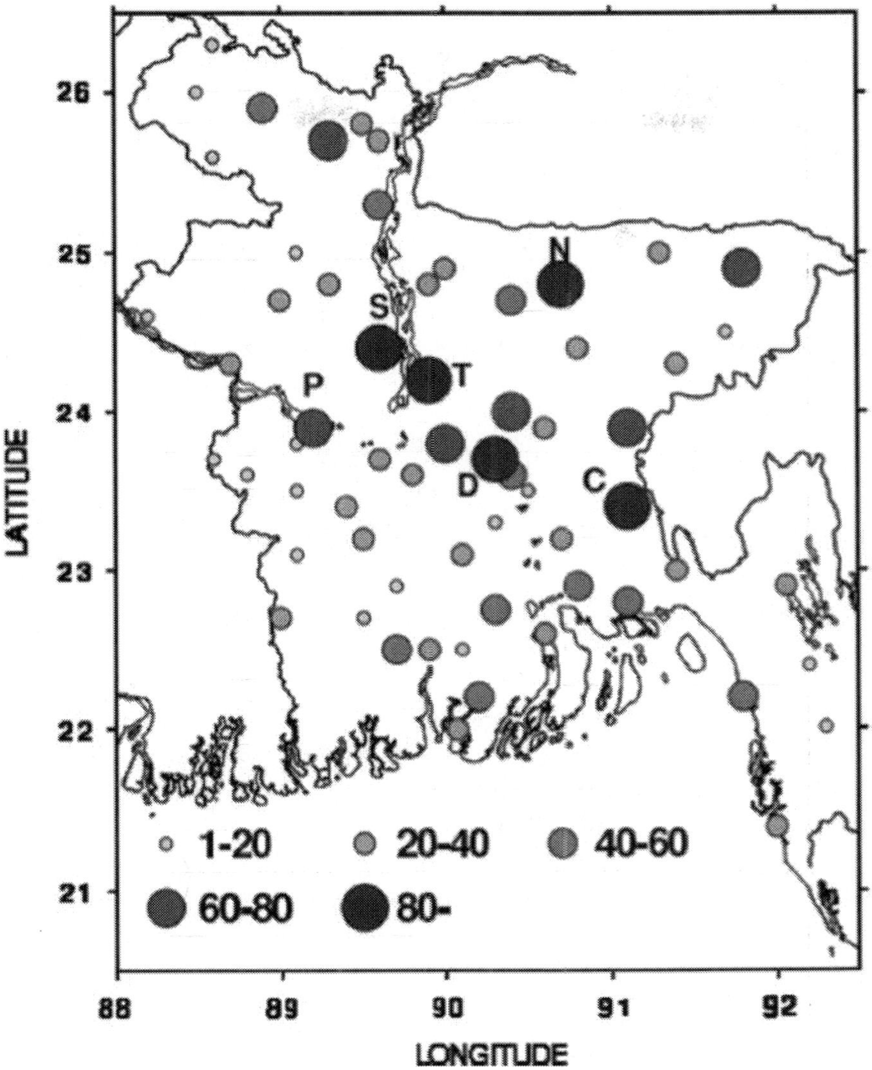

Figure 4. Geographical distribution of tornadoes in Bangladesh from 1990 to 2005. *Source:* Yamane et al., 2009.

The inhabitants of the coastal zone used to adapt and manage the frequent storms and cyclones using their wisdoms usually building their homes on raised floor, low height and surrounded by highly protective windbreaks with coconut, areca nut and fishtail palm etc. (Fig. 5).

Figure 5. Homes with windbreaks of the coastal zone.

Hill forests of the eastern and northern region and the central Sal forests are the natural defense system used to protect the human habitats from storms and tornadoes, erosion and land-slides Traditionally, the homes of the 1960s or before, people used to choose multipurpose tree and shrubby plants specially to serve the purpose of wind protection, erosion control and habitat restoration in addition to uses for health conscious drinks and foods (Rahman, 2004a and Rahman, 2004b). Maintaining their appropriate spacing, coconut, areca nut, dates, Palmyra palm, fish-tail palm, Cyperus, and Screw pine are suitable species in this respect, and also offer natural protection to shelter a home from the adverse condition. Palms can withstand 350 km/h wind speed and their specially designed highly elastic leaves can break the wind speed. These plants altogether provide a triple-tier protection measure against tidal surges and strong winds (Rahman, 2009, Rahman, 2010 and Morton, 1988).

The houses of the coastal zone are usually made on wooden platforms on raised earth excavating two or three ponds in each and every home. The flood plains were nicely conserved by digging deep ponds for multipurpose uses. Excavated earth used to meet the purpose of house building on raised landscape and the houses were designed to face the south to get maximum sunlight (Fig. 5). Large ponds with an area above 1 acre are termed as Dighi; above 10 acres is Sagor and smaller ponds of area less than an acre are called Pukurs which are used as water reservoir for domestic uses, irrigation and swimming and also for fish, shrimps, crabs and duck culture.

The houses of the coastal hilly areas of Chittagong and Chittagong Hill Tracts are traditionally designed with a central king post with suspended roof to protect from the earthquakes. Still in Chittagong and Cox's Bazar, there are Khyangs, the Buddhist temples with traditional king posts usually with large tree trunk erected at the center of the temple for load sharing during the jolting of earthquake. Over thousands of years, people living in various geographical and climatic regions of the world have evolved diverse indigenous regimes of rainwater harvesting and management as an adaptation to climate change. In South Asia, rainwater harvesting has a history of continuous practice over 8000 years and is still used today.

There are other traditional practices e.g. rainwater harvesting and storage in large earthen vats for domestic usages, this practice is a very common one for the coastal island-dwellers (Rahman, 2012). Traditional rainwater collection was very simple and was usually done by tying an old Saree (cloth) or a sheet to four posts in the yard and collecting the water in a traditional earthenware pot, a Motka (Asef et al. 2012).

Among the other traditional practices, people of the coastal zone used to do procure their necessary goods from the Haats and Bazars (markets) during the daytime to avoid the transportation problems at night.

4.1. Traditional landscaping for homes in the floodplains

Typically, rural homesteads follow a courtyard layout. The basic features are: a group of separate structures surround an open space which is defined as the courtyard. Each structure is essentially a one-roomed accommodating different function such as dwelling units for extended family members, kitchens and granaries. Toilets and outhouses such as cowsheds are located on the periphery of the homestead and where recycling of biomass takes place. The layout is introverted, that is, the buildings face away from the outside and are accessed through the courtyard. Entry into the compound from the outside is through gaps between structures. The homestead is extensively planted with trees along the boundary which strengthens the introverted layout. In flood-prone areas, an indigenous practice is to build homesteads on a raised mound, built with earth from the excavation of canals and ponds. Presently, because of resource constraints, it is not always possible for people to raise homesteads adequately above flood level. Planting design and selection of trees is the main aspect of landscaping. Land selection, preparation and homesteads are also part of landscaping. Homestead cropping maintains special adaptation procedure for raising creeper and climber crops on raised mound made with earth boulders to prevent water-logging in the wet monsoon. This is a scientific process to break the capillary system by the air-spaces among the earth-boulders (Rahman, 2004a, Rahman, 2004b, APDC, 2005 and Ahmed, 2006).

Among the other traditional adaptation practices, the season-based cropping to maintain the flora and fauna of soils of dry and wet phases; floating agriculture and deepwater rice cultivation are also important (Word Food Programme). People used to aware of getting the indication of storms and winds watching the mouth of the nests of the weaving birds (Peppler, 2010).

5. MAL-ADAPTATION PRACTICES AND DISASTERS

Cyclone Gorki-1970, Bangladesh Cyclone-1991, Sidr-2007, Nargis-2008, Aila-2009 and Mohasen-2013 have caused significant damages of life and properties and resources. From the Table 1, it is clear that, Bangladesh has achieved great progress in awareness and minimized the deaths significantly but considering the overall property and resource loss, the proportionate loss is significantly higher in the recent years (BARCIK Report, 2013). It has been observed that, unplanned afforestation, social forestry, shelterbelts, embankments and destruction of mangroves for ship-breaking and shrimp-culture; excavation of shallow ponds pisciculture and so forth are the major causes of increased destruction of properties and resources in the recent years. Huge excavated shallow ponds get heated and create low pressure centres; thus increase the frequencies of local tornadoes and cyclones (Rahman, 2012).

5.1. Conventional methods and policies for reducing calamities

Embankments, polderization, coastal afforestation and shelterbelts, construction of shelter-house are important activities. But in most cases these are constructed against the natural forces, unscientifically and inefficiently. Embankment and polders have caused permanent water-logging in many parts of the coastal zones. Many coastal polders, constructed to protect agricultural land from saltwater inundation, were subsequently turned into large shrimp farms. Saltwater was allowed into the polders in order to raise shrimps. Driven by commercial interests, the land used for agriculture and mangroves was converted, often forcibly to shrimp farming (Haque, 2004) leading to many land-use conflicts (Karim, 2004 and Deb, 1998), environmental pollution and social unrest (Firoze, 2003, Rahman, 2011b and Miyan, 2012).

Shrimp hatcheries pose a great threat to coast line. Cox's Bazar, the world's longest sea beach, is under threat of erosion due to disposal of effluents from the shrimp hatcheries (Fig. 6). Moreover, expansion of the sea beach hotels, motels and recreation zone by cutting shoreline hills has destroyed a great part of the beach which needs immediate attention to prevent further destruction (Rahman, 2010, Rahman, 2011a and Miyan, 2012). Recently, the government took action to protect the sea beach with sand in jutex bags (Fig. 7).

Figure 6. Coastal erosion due to shrimp hatcheries in Cox's Bazar and erosion control measure by sand jutex.

Figure 7. Coastal erosion due to shrimp hatcheries in Cox's Bazar and erosion control measure by sand jutex.

5.1.1. Coastal green belt

Dense forests can attenuate wave velocity (Mascarenhas, 2006). To protect local, coastal and regional areas from storms, cyclones, tornadoes and tidal upsurges, it is essential to reduce the wind speeds by planting appropriate tree species which can withstand the high speed wind and break the wind speeds. Naturally grown halophytic plants such as Sundri (*Heritiera minor*), Geoa (*E. agallocha*), Goran (*Ceriops* sp.), Kankra (*Bruguiera gymnorhiza*), Khamo (*Rhizophora mucronata*), Baen (*Avicennia officinalis*), Keora (*Sonneratia apetala*) and Kulsi (*Aegiceros majus*) have the special adaptation for withstanding in the littoral zones with clayey alluvial soil, tides and strong salinity and winds (FAO Corporate Document Repository, and Dutta, 1998). There are several palm species for instance, Golpata (*Nipa fruticans*), Hital (*Phoenix paludosa*), coconut (*Cocos nucifera*) and cane (*Calamus tenuis*) and some swamp elephant grasses (*Typha angustata* and *Typha elephantiana*), *Alpinia allughas* and screw pine (*Pandanus facicularis*) have the soil-binding capacity and control erosion. The palms also reduce the speed of tidal upsurges. Moreover, palms maintain a gradual hierarchy and reduce the speed of the strong winds. Most of the palms can withstand winds at a speed more than 250 km/h. Palms can easily break strong wind-flow (Rahman, 2012).

In order to protect life better in coastal areas from cyclones and tidal surges, in 1966 the Forest Department began a mangrove plantation program outside the protective coastal embankments. The program was based on evidence that the Sundarbans natural mangrove forests provided effective protection from wind and waves for the western coastal areas and, thus, it was expected that the plantation of mangrove would give a worthwhile degree of protection to other coastal areas of the country open to the sea. The Coastal Afforestation Scheme had been operated from 1966 to 1974, and established 4745 ha of plantations; the project was extended to 1980, by which time another 29,700 ha were raised. All these plantations were concentrated on the offshore islands and new accretions. The afforestation species were Keora (*S. apetala*), Baen (*A. officinalis*), Kankra (*B. gymnorrhiza*), Golpata (*N. fruticans*) and Gewa (*E. agallocha*). Of late, the coastal green belt has been undergoing indiscriminate destruction and encroachment. A program of participatory mangrove plantation involving nearby coastal communities, which has proved successful in other countries, could be a sustainable mechanism to protect mangrove forests. This mechanism is being actively considered within the Forest Department. However, still there lie problems with the selection of species. It has been noted that the species planted along the coastal embankment are not according to their habit and habitats, on many occasions the damage during the cyclones increased because inappropriate of choice of species (Rahman, 2011a).

Thick shelterbelts on the embankments and along the roads and highways although they protect the coastal zone they also cause adverse effect by increasing wind speed as per Pascal's law which pass through the passages at high speed (Fig. 8). In both Sidr and Aila this effect increased the wind-speed of the cyclones and caused more damages in the interior region especially where shallow water-bodies for shrimp culture were made by removing mangrove forests (Rahman, 2012).

Figure 8. Thick windbreaks in the coastal embankments.

5.2. Wrong selection of species and methods of afforestation

Massive plantation was done without paying attention to the habit and habitats requirements and disrespecting natural adaptation processes. Millions of exotic trees, especially, the *Acacia* spp., Raintree (*Albizia saman*), Mahogany (*Swietenia macrophyla*), Royal Sirish (*Albizia richardiana*) and *Eucalyptus* spp. were planted in the coastal embankments and homesteads replacing the indigenous species. The optimum spacing was not followed during the plantation and the safe distances were not maintained. As the trees are shallow-rooted and in the high water-table zone the root system could not develop for proper anchorage and thus during the cyclonic storms and tornadoes they were uprooted damaging the infrastructures and utilities and crops. The post-Sidr study showed that about 16.84 million woody and fruit trees were uprooted by cyclone Sidr. Haq et al., 2012 claimed that due to raising saplings in the earthen pots the taproot system could not develop which is found to be contradictory in field observations as it is solely an adaptive practice of habit and habitats of the plants (Rahman, 2012). The so-called Sorjan planting system is not a new practice as the people traditionally excavated several ponds in every home for landscaping to make their homes and to grow trees on raised lands.

It was also reported that due to the so-called development of land following Indonesian model ridges and furrows method, the coastal plantations were affected but it was also contradictory to traditional land management by excavating 3 or 4 deep ponds in each house (Rahman, 2012), for hundreds of years back (Figs. 5–7) (Rahman, 2012).

Plants absorb millions of liters of water, everyday and about 95% of water transpire to the atmosphere using 5% only (Verma, 1999 and Rahman, 2013a) and thus the plants keeps the atmosphere humid and maintain the hydrological cycle through transpiration and precipitation. Physiological studies revealed that Eucalyptus as a xerophytic plant has a low transpiration rate and controls stomatal opening according to water availability (Kumar, 2004, Brown et al., 2004, Ackerson, 2004 and Singh et al., 2004). Similarly comparisons have been done among xerophytic *Acacia auriculiformis, Acacia magium* and native *Pterocarpus indicus* and observed that the stomatal size is almost half in case of *A. auriculuformis and A. mangium* than that of *P. indicus* (Combalicer et al., 2012). Verma (1999) also reported very low transpiration of *Acacia* spp. Significant reduction of stomatal transpiration during dry period in *Eucalyptus* spp, *A. auriculiformis and A. mangium* were reported by many scientists (Whitehead and Beadleb, 2004, Xiaojing et al., 2008, Rahman et al., 2013, Ali et al., 2009 and Lima et al., 2003). Thus it is proved that these exotic species provide less transpiration and impacts upon humidity and low precipitation in the region and thus huge Eucalyptus and Acacia spp. are causing dryness and changing the climate. Moreover, these exotic species do not provide food and shelter for birds and wild-lives and do not favor undergrowths.

5.2.1. Destruction of mangrove forest, hills and hillocks

Since 1980s the mangroves of the South and South-western coast has been indiscriminately destroyed and the land was excavated for shrimp and fishery projects. These shallow large ponds holding saline water raise the local temperature and create pockets for low pressure depression and it is one of the major causes of huge damages of Sidr-2007 and Aila-2009.

The Sitakund mangrove forest has totally been destroyed for ship-breaking yard and caused havoc with pollutants destroying the coasta biodiversity. Due to availability of cheap laborers, the ship-breaking industry is growing very fast in the coastal zone mainly concentrated in Sitakund, Baroawlia, Bhatiari and Kumira, just north of Chittagong city on the Bay of Bengal. Ship-breaking activities present both challenges and opportunities for coastal zone management. In Bangladesh, usually ship-breaking and handling of parts and spares are done manually on open beaches without considering safe and environment-friendly waste management practices although ships contain hazardous materials such as lead, cadmium, organo-tins, arsenic, zinc, chromium, oils, organophosphates and asbestos etc. which cause cancer to lung, skin, intestine, kidney, liver or bladder and damage blood vessels (YPSA, 2005, YPSA, 2010 and Rahman, 2011a).

A ship-breaking activity is a threat to both the terrestrial and marine environment as well as to public health. It is like a mini-version of a city that discharges every kind of pollutants a metropolis can generate like liquid, metal, gaseous and solid pollutants. Oil films on water reduce the exchange of oxygen and carbon dioxide across the air-sea interface which is harmful to aquatic life. Critical concentration of dissolved oxygen (DO) and higher biological oxygen demand (BOD) were found with an abundance of floatable materials: grease balls and oil films in the seawater (UK Essays, 2013). It also causes damage to the bird population by coating their feathers with oil which causes buoyancy and insulation losses. Sometimes spilling may cause wide spread mortality amongst the population of fish, mammals, worms, crabs, mollusks and other water organisms. Furthermore, oil spilling may cause serious damage by reduction of light intensity, inhibiting the exchange of oxygen and carbon dioxide across the air-sea water interface, and by acute toxicity. As a result the growth and abundance of marine organisms especially plankton and fishes may be seriously affected. Indiscriminate expansion of ship-breaking activities poses a real threat to the coastal inter-tidal zone and its habitat (YPSA, 2005, YPSA, 2010, Rahman, 2011a and Hossain and Ahmed, 2008).

Moreover, cutting of hill and hillocks for industrial establishment in Chittagong is causing a great land area loss and exposure of the rocky strata. Some folded ranges in Cox's Bazar are being cut for residential accommodation. The coast line of Cox's Bazaar has been dug for shrimp-culture which will in the long run make the coastal belt vulnerable to sea current erosion. Hill-cutting may cause land-slides, faults during earth quake.

Unplanned development of resort area after cutting the hills and establishment of shrimp hatcheries along the coast threatens the beach with erosion, while waste from tourists and vendors are damaging the aquatic habitat. About 15 km of the Cox's Bazar beach is at high risk of erosion, especially the shrimp hatcheries at Himchari and Kolatoli, and also motel and hotel zone, and the Diabetic Hospital areas of Kolatoli. Half of the Kolatoli village has already been lost to erosion due to unplanned discharges of water to the sea from 55 shrimp hatcheries (The Daily Ajker Deshbidesh, 2008). Sea current has damaged 3.4 km of beach of the Old Marine Drive Road from Kolatoli to Himchari in Cox's Bazar (Rahman, 2011a).

6. PURVIEW OF POLICY CHALLENGES

Despite increasing recognition of the need for Integrated Coastal Zone Management (ICZM) strategies, individual coastal policies in Bangladesh are still mainly formulated with narrow sector-oriented objectives in mind and are not part of an overall framework. Considering the following three reasons the Government of Bangladesh initiated the coastal zone policy in 2005 focusing on the fact that: (a) the coastal zone is lagging behind in socio-economic developments on many aspects; (b) There are poor initiatives to cope with different disasters and gradual deterioration of the environment; (c) the coastal zone has the potential to contribute much to national development.

Under Coastal Zone Policy it is noted that "for a Strategic Planning and Program Development, a Coastal Development Strategy for poverty reduction, economic growth and social development will be formulated and implemented". But the policy documents produced by the government did not consider proper participation of natural resource dependent occupational groups and thus it recommends educating local people on climate change issues as an external matter. In the policy document there is a priority area identified to conduct research on locally adapted crop varieties but it failed to respond to local farmers' knowledge of crop varieties and there is no clear indication of farmers' participation in the research activities. These policy documents are costly to produce yet they end up not being implemented (BARCIK Report, 2013).

The strategy should have a time and resource-bound specification with priority actions in coastal development. It should be about building a process to implement the policies, not preparing a classic master plan. Such a strategy makes critical choices, for example in relation to targeted regions, disadvantaged groups and issues. Focus should be on implementation, including a set of indicators and corresponding monitoring arrangements to assess performance. The strategy document should reflect the specific actions needed to achieve coastal development objectives. In particular, links should be made to the content and process of the National Strategy for Economic Growth and Poverty Reduction and the subsequent multi-year development plans and to other national policy and planning processes. The strategy document should be a component of the chain: Policy⇒strategy⇒priority investment Program (Coastal Zone Policy, 2005).

Similarly, although many sectoral policies have clear implications for coastal development, in most cases they do not have specific sections on coastal areas and often fail to capture the distinctive combinations of vulnerabilities and opportunities that characterize the coast. The coast remains an area of institutional weakness (World Bank Report, 2013). Though several government agencies and NGOs are working in the coastal region, there are limited linkages between them and institutional fragmentation is common. Further, many of the government agencies responsible for coastal policies have hardly any presence at the local level. Insufficient coordination—compounded by a lack of institutional, financial, and human capacities for implementation and monitoring impedes effective policy action. Consequently, sectoral development policies, the National Adaptation Programme of Action (NAPA), and the Bangladesh Climate Change Strategy and Action Plan (BCCSAP) all suffer from inadequate execution and follow-up. Too often, policy efforts aim at formulating master plans rather than at creating flexible planning and implementation procedures that can be adapted to changing situations and priorities based on performance monitoring and continuous feedback from stakeholders (Michel and Pandya, 2010). Decisions for policy making and implementation usually go from top to bottom; due to this, various developments of disasters such as cyclone Sidr- 2007 and cyclone Aila- 2009 have been witnessed in Bangladesh. Therefore, a bottom up approach involving the grassroots voice, integration of local and indigenous knowledge should be incorporated to combat hydro-meteorological hazards and other risks (BARCIK Report, 2013).

Practice and experience developed over the years suggest that successful ICZM strategies should be in accordance with nationally accepted criteria for the development of the coastal zone, taking into account poverty alleviation and economic growth, environmental quality, empowerment of local communities and local participation in planning and implementation, disaster preparedness and mitigation, and international cooperation. It should stimulate and facilitate operational interactions among agencies needed to implement sector policies and create a national platform to facilitate implementation and monitor corresponding progress. Most importantly, the scientific values of the traditional knowledge should be evaluated and a comprehensive policy should be made through dialogue between scientists and policy makers.

The key to ICZM is to empower timely decision making at the appropriate levels, providing the flexibility to respond to evolving circumstances and cope with persistent uncertainties. When this has been achieved, the great range of opportunities offered by the coastal zone including marine resources; the accretion of new land, oil and gas resources; the potential productivity of mangroves and other coastal ecosystems; and tourism and leisure activities can be judiciously developed. However, there is inadequate information on pollution and sedimentation loads entering the Bay of Bengal and impacts on coastal habitats and endangered species. This information is necessary in order to understand the functioning of the ecosystem and its reaction to stress over time (NOAA, 2008).

The initiation of the Coastal Zone Policy and ICZM has not contributed any significant improvement rather overall coastal ecosystems have been rapidly deteriorated. The mangroves are diminishing and the effect of cyclones, floods and tidal surges are frequent. The intensity of damages of properties has greatly increased. Both biodiversity and human life have been affected due to lack of scientific information and exclusion of traditional adaptation practices (BARCIK Report, 2013). Bangladesh Resource Centre for Indigenous knowledge (BARCIK) also studied on NAPA and BCCSAP and found that an amount of BDT 3.0 billion (USD 38.46 million) was allocated in 2008–2009 to deal with impacts and adaptation of climate change but nothing was spent during that period. Later, government finalized the BCCSAP in the light of government's Vision 2021 and perspective action plan and allocated BDT 7.0 billion (USD 89.74 million) to Climate Change Fund in 2009–2010 fiscal year. Reports also pointed out that for every project, budget and project design cost has been proposed where more experts in the preparation of details project will also be included for donor hunting as loan or grant. The NAPA document has ignored the communities understanding, their experiences and observations, and the traditional and indigenous knowledge and practices. In every individual project documents, the writing teams have considered to involving several experts, infrastructural constructions, new project formulation but have not addressed how the existing similar projects could be incorporated and coordinated. That actually results in 'action when the disaster starts'. Virtually very little progress has been achieved to mitigate the long-term climate change impacts other than minimizing human death.

The major challenge to ICZM is to realize these potentials while mitigating or adapting to vulnerabilities through a process that enhances the livelihoods of the inhabitants and provides communities with avenues for input to, and support from, external institutions and of course traditional adaptation practices and the natural and traditional defense mechanisms.

For Bangladesh, the political institutional barrier is the largest obstacle to overcome, requiring awareness and capacity building at the highest political and policy levels. Increased funding would help, but must be channeled toward the most climate-vulnerable communities through appropriate structures to maintain accountability and transparency. A coherent, dynamic national action plan is needed to prioritize adaptation measures, regularly review implementation, and eventually revise and reprioritize policy actions. Such a plan should facilitate more comprehensive evaluation of adaptation options, providing a framework for assessing social and environmental as well as economic costs and benefits across all relevant sectors and stakeholders. Formulating this plan would force decision makers to think beyond the short-term, politically expedient horizons of three to five years. A shift to long-term planning will enable the long-term effects of climate change—with predicted rises in sea levels, possible increases in the frequency of major storms, and changes in rainfall patterns over the whole Ganges–Brahmaputra basin—to be taken into account.

For an effective ICZM, natural phenomenon of wind and water flow and the interaction with the adaptation procedures of biological systems must be considered. Protection of coastal lands, forests, shorelines, water-bodies and islands as well as land accretion such as charlands, sand-dunes and reduction of effects of climate extremes especially the tropical cyclones and tidal surges, the laws of hydrodynamics and wind-forces and the interaction of biological adaptations especially of the trees must be considered and utilized for a sustainable coastal zone management.

7. CONCLUSIONS

The above synthesis indicates that, the traditional natural and artificial defense mechanisms are vital to protect the flood plains and the whole coastal zones. However, introduction of some impracticable and unsustainable practices have almost destroyed the long-term adapted cultures. The floodplains can be managed by landscaping and planting appropriate species for protection of the structures and lives from natural calamities like cyclones, tornadoes, floods and tidal surges. Hills and wet-bodies are also important natural protection measures and provide adaptation for diverse species. Since coastal zone of Bangladesh is very vulnerable to extreme events, there is no short-cut solution to protect the region from the vulnerability except by investing in natural and traditional defense mechanism. Conventionally, it was tried with many practices like polderization, embankments, dykes and improvement of roads and highways but these are only increasing the sufferings of people and increasing economic loss causing huge migration to the cities. In the name of "Grow More Food", many wet-bodies were filled up with earth but lost the natural catchments for holding surface water and people are now almost dependent on ground water. Thus, severe water crisis has been emerging. Floods too have become frequent damaging huge crops and livelihoods every year. Massive plantation of some of the exotic and xerophytic trees are driving the country towards desertification and tree food and feed crisis thus losing biodiversity. The policy makers, as well as the researchers are required to come up with better practices for an improved coastal zone management. Countries like India, Myanmar, Cambodia and Vietnam where coastal zones are very vulnerable to extreme events need more interaction and integration of the traditional ways of adaptation. The natural and traditional adaptation practices should be evaluated scientifically through habitat research, interaction with communities and in-depth analysis in the workshops and seminars involving stakeholders for a sustainable national and regional policy for coastal zone management.

ACKNOWLEDGEMENTS

The study was conducted with the financial support received from the Asia Pacific Networks for Global Change Research (APNGCR) project ARCP2011 16NMY-IGBP through the International Geosphere and

Biosphere Programme (IGBP). The Research Team thankfully acknowledges their support. Sincere thanks go to Dr Opha Pauline Dube, Dr Mannava Sivakumar and to Professor M Alimullah Miyan for their coordination and guidance. The Authors sincerely acknowledge IUBAT—International University of Business Agriculture and Technology for giving the opportunity to do this study on Natural and Traditional Defense Mechanisms in Coastal Zones of Bangladesh which is a very import issue at present.

REFERENCES

1. Ackerson, 2004. Eucalyptus Enduring Myths, Stunning Realities . In: Abbasi, S.A. Ramenh, N., Vinithan, S. (Eds.). Discovery Publishing House 4831/24, Ansari road, New Delhi 110002, India ISBN W8171418929. pp 26-27.
2. Afroz, R., Rahman, M.A., 2013. Transboundary river water for ganges and teesta rivers in Bangladesh: an assessment. Global Sci. Technol. J. 1, 100–111.
3. Ahmed, K. Iftekhar., 2006. The rural Bangladeshi courtyard. BRAC Univ. J. 3, 9–15. Ali, A., 1999. Climate change impacts and adaptation assessment in Bangladesh.Clim. Res. 12, 109–116.
4. Ali, I., Abbas, S.Q., Hameed, M., Naz, N., Zafar, S., Kanwal, S., 2009. Leaf anatomical adaptations in some exotic species of eucalyptus l'hér. (Myrtaceae). Pak. J. Bot. 41, 2717–2727.
5. APDC, 2005. Handbook on Design and Construction of Housing for Flood-prone Rural Areas of Bangladesh 984-32-2163-X. Nymphea, 67/D, Panthapath, Dhaka-1209, Bangladesh; ⟨ http://www.adpc.net/audmp/library/housinghandbook/ handbook_complete-b.pdf⟩ .
6. Asef, M., Masum, S.M.M., Paul, S., 2012. Investigation of Utilizing Rainwater as Alternate Source of Water in Tejgaon Industrial Area Thesis; Bachelor of Civil Engineering. Ahsania University of Science and Technology, Tejgaon, Dhaka Bangladesh p. 2012 ⟨ https://www.scribd.com/doc/ 102429752/Investigation-o

7. f-Utilizing-Rainwater-as-Alternate-Source-of-Water-in-Tejgaon-Industrial-Ar ea-Main-Signed⟩ (submitted in May).
8. Bangladesh State of The Environment, 2001. ⟨http://www .moef.gov.bd/html/state_ of_env/pdf/bangladesh_disasters.pdf⟩ (accessed on 11 July 11, 2014).
9. Banica, A., Bastard, J., Kosiek, M., 2003. Integrated Coastal Zone Management (ICZM): A Framework To Tackle Environmental Issues? Danish Approach Centre for Environmental Studies, June 2003 Reprocenteret, Det Naturvidenskabelige Fakultet, Aarhus Universitet, (2003) 87-7785-146-3.
10. BARCIK Report, 2013. National Adaptation Programme of Action (NAPA) and People's Adaptation Plan. ⟨http://www.klima-und gerechtigkeit.de/fileadmin/ upload/Dialogforen/summary_NAPA_study.pdf⟩.
11. Bhattacharya, S., Chattopadhyay, D., Mukhopadhyay, A., 2013. Changing dimensions of food security in a globalized world: a review of the perspectives for environment, economy and health. Int. Res. J. Environ. Sci 2, 67–73.
12. Bird, E.C.F., 1979. Environmental problems related to the coastal dynamics of humid tropical deltas; In: Proceedings of the Jakarta Workshop on Coastal Resources Management, ⟨http://www.unu.edu/unupress/unupbooks/80130e/80130E04. htm⟩.
13. Biswas, K.K., Emmelin, L., 2010. Changing Climate; Bangladesh Facing the Challenge of Severe Flood Problems; A Comparison of Flood Management between Bangladesh and the Netherlands ⟨http://www.bth.se/fou/cuppsats.nsf/all/ 4ac7b 480ad562fb0c1257749005fc03e/$file/Thesis2010.pdf⟩.
14. Brammer, H., 1990. Floods in Bangladesh: geographical background to the 1987 and 1988 floods. Geogr. J. 156, 12–22.
15. Brown et al., 2004. In Eucalyptus Enduring Myths, Stunning Realities. Abbasi S.A., Ramesh, N., Vinithan, S., (Eds.), Discovery Publishing House 4831/24, Ansari road, New Delhi 110002, India, ISBN W8171418929, pp 26–27.

16. BWDB, 2013. Coastal Embankment Improvement Project Phase-I, Environmental Impact Assessment of Polder 35/3; May, 2013, Government of the People's Republic of Bangladesh, Ministry of Water Resources, Bangladesh Water Development Board 〈http://www-wds.worldbank.org/external/default/WDSContentServer/WDSP/IB/2013/08/28/000356161_20130828151210/Rendered/INDEX/E41410v40CEIP0000PUBLIC00Box379813B.txt〉.
17. Chand, B.K., Trivedi, R.K., Dubey, S.K., Beg, M.M., 2012. Aquaculture in Changing Climate of Sundarban: Survey Report on Climate Change Vulnerabilities, Aquaculture Practices & Coping Measures in Sagar and Basanti Blocks of Indian Sundarban, West Bengal University of Animal & Fishery Sciences, Kolkata, India.
18. 〈http://www.nicraicar.in/nicrarevised/images/Books/Aquaculture%20in% 20Changing% 20Climate%20of%20Sundarban.pdf〉 (accessed on 11 July 11, 2014).
19. Climate Change Cell, 2007. Bangladesh and Climate Change, Climate Change Cell, Department of Environment, Government of the People's Republic of Bangladesh 〈http://www.bdresearch.org.bd/home/climate_knowledge/cd1/pdf/Bangladesh%20and%20climate%20change/Climate%20change%20impacts%20,vulnerability,%20risk/Climate%20Change%20And%20Bangladesh.pdf〉.
20. Climate Change Cell, 2009. Climate Change, Gender and Vulnerable Groups in Bangladesh. Climate Cell, DoE, MoEF; Component 4b, CDMP, MoFDM. Dhaka.
21. Coastal Zone Policy, 2005. Ministry of Water Resources; Government of the People's Republic of Bangladesh. 〈http://www.warpo.gov.bd/pdf/czpo_eng.pdf〉.
22. Combalicer, M.S., Lee, D.K., Woo, S.Y., Hyun, J.O., Park, Y.D., Lee, Y.K., Edwin, A., Combalicer, E.A., Tolentino Jr., E.L., 2012. Physiological characteristics of Acacia auriculiformis A. Cunn. ex Benth., Acacia mangium Willd. and Pterocarpus indicus Willd. in

the La Mesa Watershed and Mt. Makiling, Philippines. J. Environ. Sci. Manage. Spec Issue 1, 14–28.

23. Cruz, R.V., Harasawa, H., Lal, M., Wu, S., Anokhin, Y., Punsalmaa, B., Honda, Y., Jafari, M., Li. C., Huu, N., 2007. Asia Climate change, 2007, impacts, adaptation and vulnerability. In: Parry, M.L., Canziani, O.F., Palutik of, JP, et al. (Eds.), Contribution of Working Group II to the Fourth Assessment Report of the Intergovernmental Panel on Climate Change. Cambridge University Press, Cambridge, pp. 469–506.

24. Deb, A.K., 1998. Fake blue revolution: environmental and socio-economic impacts of shrimp culture in the coastal areas of Bangladesh. Ocean Coast. Manage. 41, 63–88.

25. DLS, 2013., Training Guide on Disaster and Climate Risk Management in Livestock Sector Project for Officers Training of DLS on DRR & CCA ⟨http://www.dls.gov.bd/Training%20Guideline%20for%20disaster%2021.03.2013.pdf⟩.

26. DoE Publication., 2013. ⟨http://www.doe.gov.bd/publication_images/17_environ ment.pdf⟩.

27. Dube, S.K., 2009. Recent Developments in Storm Surge Prediction in the North Indian Ocean ⟨http://nidm.gov.in/idmc2/PDF/Abstracts/Cyclone.pdf⟩.

28. Dutta, A.C., 1998. Botany for Degree Students. Oxford University Press, New Delhi, India, ISBN: 9780195637489, pp. 324–330.

29. Earth Justice ACJP, 2009. The Role of Black Carbon in Endangering World Heritage Sites Threatened by Glacial Melt and Sea Level Rise.

30. FAO Corporate Document Repository; Mangrove Trees and Shrubs ⟨http://www.fao. org/docrep/010/ai387e/AI387E06.htm⟩.

31. FAO, 2014. Document Repository: Sustainable Management of the Natural Resource Base. Natural Resources Management and Environment Department, FAO, Rome ⟨http://www.fao.org/docrep/t3384e/t3384e03.htm⟩.

32. Firoze, A., 2003. The Southwest Coastal Region: Problems and Potentials. The Daily Star, XIV, Dhaka (Issue 215).
33. Flood Archive., 2004. Global Register of Major Flood Events. ⟨http://www.dart mouth.edu/ floods/Archives/ 2004sum.htm⟩.
34. GDRC, 2014. Things you Need to Know of Small Island Developing States: An Overview of Major Environmental Threats Encountered by Small Island Developing States. ⟨http://www.gdrc.org/oceans/10things/⟩.
35. Geotech Guideline., 2005. Guidelines on Erosion Control and Drainage of Railway Formation (Guideline No. GE: G-4) February- 2005 Geo-technical Engineering Directorate, Research Designs and Standards Organisation, Lucknow-226011 ⟨http://www.vetiver.com/IND_Indian%20railways.pdf⟩.
36. Global Humanitarian Forum, 2009. Climate Change—The Anatomy of a Silent Crisis: Case Study: Bangladesh a Nation at the Frontline of the Climate Change Crisis, Human Impact Report Climate Change Global humanitarian Forum Geneva, ⟨http://www.ghf-ge.org/human-impact-report.pdf⟩.
37. Government of Bangladesh, 2008. Bangladesh Climate Change Strategy And Action Plan. Government of Bangladesh. ⟨http://www.foshol.org/IUCN_BCCSAP_2008/ IUCN_BCCSAP_2008_part2.pdf⟩.
38. Haq, M.Z., Robbani, M., Ali, M., Hasan, M.M., Uddin, Begum da Silva, M.J., Pan, X.Y., Karim, J.A.T., 2012. Damage and Management of Cyclone Sidr—Affected Home- stead Tree Plantations: A Case Study from Patuakhali. Springer, Bangladesh (Scienceþ Business Media B.V.).
39. Haque, A.K.E., 2004. Sanitary and Phyto-Sanitary Barriers to Trade and its Impact on the Environment: The Case of Shrimp Farming in Bangladesh. Dhaka, IUCN Bangladesh, Country Office, 63 pp.
40. Hossain, M.K., Ahmed, F.U., 2008. Nursery Manual of Endangered Tree Species. Arannyak Foundation, Banani, Dhaka.

41. Hossain, F., 2002. Floods 2002. The Associated Press 〈http://www.greatdreams.com/ floods_2002.htm〉.
42. IMD, 1979. Tracks of Storms And Depressions in the Bay of Bengal and the Arabian Sea 1877–1970. India Meteorological Department (IMD), New Delhi.
43. IPCC, 2001. Coastal zones and marine ecosystems. In: McCarthy, J.J., Canziani, O.F., Leary, N.A., Dokken, D.J., White, K.S. (Eds.), Climate Change 2001: Impacts, Adaptation, and Vulnerability. Cambridge University Press, Cambridge.
44. IPCC, 2007. Oceans and Coastal Zones Climate Change: Working Group II: Impacts, Adaptation and Vulnerability. 〈http://www.ipcc.ch/publications_and_data/ar4/ wg2/en/ch10s10-2-4-3.html〉.
45. Karim, N., 2004. Options for Floods and Drought Preparedness in Bangladesh. 〈www.grif.umontreal.ca/pages/papers 2004/paper-karim N pdf〉.
46. Khan, M.S., Rahman, M.M., Ali, M.A., 2001. Bangladesh National Herbarium, Red Data Book of Vascular Plants Bangladesh, Bangladesh National Herbarium.
47. Khan, T.A., 2000. Water based disasters in Bangladesh. In: Q.K. Ahmad (Ed.), Bangladesh Water Vision 2025: Towards a Sustainable Water World. Bangla- desh Water Partnership (BWP), Dhaka. pp. 54-62.
48. Kumar, D., 2004. In: Eucalyptus Enduring Myths, Stunning Realities. Abbasi, S.A., Ramesh, N., Vinithan, S. (Eds.). Discovery Publishing House 4831/24, Ansari road, New Delhi 110002, India, ISBN W8171418929, pp 26–27.
49. Lima, W., de, P., Jarvis, P., Rhizopoulou, S., 2003. Stomatal responses of eucalyptus species to elevated CO2 concentration and drought stress. Sci. Agric. 60, 231–238.
50. Mascarenhas, A., 2006. Extreme events, intrinsic landforms and humankind: post- tsunami scenario along Nagore-Velankanni Coast, Tamilnadu. Curr. Sci. India 90, 1195–1201.

51. Michel, D., Pandya, A., 2010. Coastal Zones and Climate Change; Stimson's Regional Voices, ISBN: 978-0-9821935-5-6. ⟨www.stimson.org/programs/regional-voices⟩.

52. Mirza, M.M.Q., 2002. Global warming and changes in the probability of occurrence of floods in Bangladesh and implications. Global Environ. Change 12, 127–138. Miyan, M.A., 2009. Monsoon in Bangladesh; Changes and Adaptation: Asian

53. Monsoon Years AMY-6, 30 November–1 December, 2009. Kunming China,

54. Miyan, M.A., 2012. Coastal Zone Management of Bangladesh, Fourth Session of the Joint WMO-IOC Technical Commission for Oceanography and Marine Meteor- ology (JCOMM-4) Yeosu, Republic of Korea.

55. MoEF., 2001. Bangladesh: State of the Environment 2001 ⟨http://www.moef.gov.bd/ html/state_of_env/pdf/bangladesh_disasters.pdf⟩ (accessed on 11 July 11, 2014).

56. Morton, J.F., 1988. The Palmyra or Toddy Palm (Borassus flabellifer L.) notes on distribution, propagation, and products of Borassus Palms (Arecaceae). Econ. Bot. 42, 420–441.

57. Nagaraju, K.N., 2012. Impact of built environment on climate change in urban centers, strategies to be adopted in reducing it. In: International Conference on Environmentally Sustainable Urban Ecosystems (ENSURE 2012), 24–26 Febru- ary 2012. Centre of Excellence for Integrated Land use Planning and Water Resource Management, Department of Civil Engineering, Indian Institute of Technology, Guwahati, Assam, India.

58. NOAA, 2008. Bay of Bengal Large Marine Ecosystem. ⟨http://www.eoearth.org/ view/article/150447/⟩.

59. NPDM, 2015. National Plan for Disaster Management 2010–2015. Disaster Manage- ment Bureau Disaster Management & Relief Division, Government of the Peoples Republic of Bangladesh.

60. OECD Report, 2003. Development and Climate Change in Bangladesh: Focus on Coastal Flooding and The Sundarbans ⟨http://www.oecd.org/env/cc/21055658.pdf⟩.

61. Pender, J.S., 2008. What Is Climate Change? And How It Will Effect Bangladesh. Briefing Paper (Final Draft). Dhaka, Bangladesh: Church of Bangladesh Social

62. Development Programme. ⟨https://www.kirkensnodhjelp.no/Documents/Kirkens%20N%C3%B8dhjelp/Publikasjoner/Temahefter/FINAL%20Draft%20WHAT%20IS%20CLIMATE%20CHANGE%20AND%20HOW%20IT%20MAY%20AFFECT%20BANGLADESH.pdf⟩.

63. Peppler, R., 2010. Fieldwork in Indian Country: A Conversational Experience. Weather and Society Watch 4, July 2010 Issue.

64. Roy, S.S., 2009. A Special analysis of extreme hourly precipitation pattern in India. Int. J. Climatol. 29, 345–355.

65. Rajeevan, M., Bhate, J., Jaswal, A.K., 2008. Analysis of variability and trends of extreme rainfall events over India using 104 years of gridded daily rainfall data. Geophys. Res. Lett. 35, L18707.

66. Rabbani, G., Rahman, A.A., Islam, N., 2010a. Climate Change and Sea Level Rise: Issues and Challenges for Coastal Communities in the Indian Ocean Region Coastal Zones and Climate Change. The Henry L. Stimson Center 978-0- 9821935-5-6.

67. Rabbani, G., Rahman, A.A., Islam, N., 2010b. Climate change and sea level rise: issues and challenges for coastal communities in the Indian Ocean Region. In: Coastal Zones and Climate Change The Henry L. Stimson Center ISBN: 978-0-9821935- 5-6 Library of Congress Control Number: 2009939156.

68. Rahman, A.A., Olayinka, B.U., Haruna, M., Yussuf, B.T., Aderemi, M.O., Kolawole, O.S., Omolokun, K.T., Aluko, T.A., Oladele, F.A., 2013. Cooling effects and humidifica- tion potentials in relation to stomatal features in some shade plants. Int. J. Appl. Sci. Tech 3, 138–152.

69. Rahman, M.A., 2004a. Model rural homestead farming— a real example of crop diversification. In: Plantation Crops and Organic Farming: Research Articles Series 1: Some Environment Related Problems and Their Solutions. ISBN 984- 32-1154-5.
70. Rahman, M.A., 2009. Palmyra Palm: Symbol for Sustainability. 〈http://feppcar.org/ 59/palmyra-palm-symbol-for-sustainability/〉.
71. Rahman, M.A., 2010. Coastal Zone Management of Bangladesh Presented in the International Geosphere-Biosphere Programme Synthesis Integration and Exploration: Global Environmental Change and Sustainable Development;
72. Needs of Least Developed Countries (IGBP-LDCs) Synthesis Workshop, Maputo Mozambique September 20–22, 2010. 〈feppcar.org/122/coastal-zone-manage- ment-in-bangladesh〉 (accessed on June 4, 2014).
73. Rahman, M.A., 2012. Workshop on Coastal Zone Management, Patuakhali University.
74. Rahman, M.A., 2011a. Study on the Changes of Coastal Zone: Chittagong to Cox's Bazar along the Bay of Bengal; Global Summit on Coastal Seas, EMECS 9 (2011), August 28–31, 2011, Baltimore, Maryland, USA.
75. Rahman, M.A., 2011b. Vulnerability of the Bay of Bengal Enclosed Coastal Sea due to Socio-Economic Conditions of the Megacity of Dhaka; Global Summit on Coastal Seas, EMECS 9, August 28–31, 2011, Baltimore Maryland, USA, Published in Indian Journal of Humanities, Vol. 1: ISSBN: 2248-9541.
76. Rahman, M.A., 2013a. Grey Water Use can Reduce Huge Water Crisis in Dhaka Megacity—〈http://feppcar.org/280/grey-water-use-can-reduce-huge-water-cri sis-in-dhaka-megacity/more-280〉.
77. Rahman, M.H., 2004b. Wetland and Water Pollution Management in Bangladesh 〈http://globalcommunitywebnet.com/gdufour/2004WorkHasiburrahman.htm〉.

78. Raygorodetsky, G., 2013. Why Traditional Knowledge Holds the Key to Climate Change 〈http://unu.edu/publications/articles /why-traditional-knowledge- holds-the-key-to-climate-change. html〉.

79. Sanjay, V.D., Balaji, V., 1994. Conservation of Mangrove Forest Genetic Resources: A Training Manual. (Eds.) Sanjay, V. Desmukh, V., Balaji. 〈http://www.mssrf.org/ csr/csr-pub/01-conservationofmangroveforestgenetic.pdf〉.

80. Sattar, M.A., Bhattacharjee, D.K., 1987. Physical and Mechanical Properties of Sundri (Heritiera fomes) and Baen (Avicennia alba). Government of the People's Republic of Bangladesh, Forest Research Institute, Timber Physics Series, Bull, Chittagong, Bangladesh p. 10.

81. Integrated coastal area management and agriculture, forestry and fisheries. FAO Guidelines. In: Scialabba, N. (Ed.), Environment and Natural Resources Service. FAO, Rome, p. 256.

82. Seal, L., Baten, M.A., 2012. Salinity Intrusion in Interior Coast: A New Challenge to Agriculture in South Central part of Bangladesh, January 2012 〈http://www.

83. unnayan.org/reports/climate/Salinity_Intrusion_in_Interior%20 Coast-A_New_Challenge_to_Agriculture_in_South_ Central_part_of_Bangladesh.pdf〉 (accessed on 9 July, 2014).

84. Sherbinin, Alex de., Carr, D., Cassels, S., Jiang, L, 2007. Population and environment. Ann. Rev. Environ. Resour 32, 345–373. http://dx.doi.org/10.1146/annurev. energy.32. 041306. 100243.

85. Singh et al., 2004. In: Eucalyptus Enduring Myths, Stunning Realities, . Abbasi SA, Ramenh N. Vinithan S. 2004; Discovery Publishing House 4831/24, Ansari road, New Delhi, 110002, India ISBN W8171418929, (1990) pp 26-27.

86. Sivakumar, M.V.K., Stefanski, R., 2011. Climate change in South Asia. In: Lal, R., Sivakumar, M.V.K., Faiz, S.M.A., M. Rahman, A.H.M., Islam, K.R. (Eds.) Climate Change and Food Security in South Asia, Springer, 600p.

87. Sultana, W., Aziz, M.A., Ahmed, F., 2008. Climate Change: Impact on Crop Production and its Coping Strategies. ⟨http://www.wamis.org/agm/meetings/ rsama08/Bari103-Sultana_Coping_Strategies.pdf⟩.
88. Sundarban Biosphere Reserve., 2006. ⟨http://www.sundarbanbiosphere.org/html_ files/sunderban_biosphere_reserve.htm⟩.
89. The Daily Ajker Deshbidesh, October 16, 2008.
90. The Daily Star, September 16, 2013. Sundarbans Shrinking ⟨http://www.thedailys tar.net/beta2/news/sundarbans-shrinking/⟩ (accessed on 16th September 2013).
91. The Sundarbans, 1997. ⟨http://whc.unesco.org/en/list/798⟩.
92. UK Essays, 2013. Bangladesh is a Test Case of Sustainability. Environmental Sciences Essay Impact of Ship Breaking on Physiochemical Properties of Seawater ⟨http://www.ukessays.com/essays/environmental-sciences/ bangladesh-is-a-test-case-of- sustainability-environmental-sciences-essay.phpftn5ixzz3IjVmL6IT⟩.
93. UNEP, 1994. Programme of Action for the Sustainable Development of Small Island Developing States, Extract from A/CONF.167/9, Part I, Annex I. ⟨http://islands. unep.ch/dsidspoa.htm⟩.
94. Verma, S.K., 1999. A Textbook of Plant Physiology and Biochemistry, third ed.1999.
95. S. Chand & Company Ltd., Ramnagar, New Delhi-110 055, pp. 85–102 (ISBN 81- 219-0627-X (1999).
96. Vos, F., Rodriguez, J., Below, R., Guha-Sapir, D., 2010. Annual Disaster Statistical Review 2009: The Numbers and Trends. Centre for Research on the Epidemiology of Disasters (CRED), Brussels.
97. Wever, L., 2008. Assessing Management Challenges and Options in the Coastal Zoneof Timor-Leste Griffith Centre for Coastal Management (GCCM) Research Report No. 86 ⟨http://databases.eucc-d.de/files/documents/00000985_ CoastalMgtET_ WeverL2008.pdf⟩.

98. Whitehead, D., Beadleb, C.L., 2004. Physiological regulation of productivity and water use in Eucalyptus: a review. For. Ecol. Manage. , http://dx.doi.org/10.1016/ j.foreco.2004.01.026 (01/2004).

99. Wong, C.M., Williams, C.E., Pittock, J., Collier, U., Schelle, P., 2007. World's Top 10 Rivers at Risk. WWF International, Gland, Switzerland.

100. World Bank Report, 2013. Warming Climate to Hit Bangladesh with Sea Level Rise, More Floods and Cyclones 〈http://www.worldbank.org/en/news/press-release/

101. 2013/06/19/warming-climate-to-hit-bangladesh-hard-with-sea-level-rise-mor e-floods-and-cyclones〉.

102. Xiaojing, L., Zhao, P., Rao, X., Ma, L., Cail, X., Zeng, X., 2008. Response of canopy stomatal conductance of Acacia mangium forest to environmental driving factors. Front. For. China , http://dx.doi.org/10.1007/s11461-008-0001-3.

103. Yamane, Y., Hayashi, T., Dewan, A.M., Akter, F., 2009. Severe local convective storms in Bangladesh: Part 1, Climatology. Atmos. Res. 95, 407–418.

104. YPSA, 2005. Workers in Ship Breaking Industries: A base line survey of Chittagong, Bangladesh. Young Power in Social Action (YPSA), Chittagong p. 79.

105. YPSA, 2010. Ship Breaking in Bangladesh: Environmental Pollution. 〈http://www. shipbreakingbd.info/ Environment. html〉.

Chapter 10

Environmental Consequences of Antarctic Tourism from a Global Perspective

Shahab Kariminia[a], Sabarinah Sh. Ahmad[a], Rugayah Hashim[b], Zulhabri Ismail[a]

[a] *Faculty of Architecture, Planning and Surveying, Universiti Teknologi MARA, Shah Alam 40450, Malaysia*

[b] *Faculty of Administrative Science and Policy Studies, Universiti Teknologi MARA, Shah Alam 40450, Malaysia*

ABSTRACT

Antarctica tourism evidently harms its ecosystem in direct and indirect ways. Ecosystems are dynamic and complex systems, which simultaneously depend on various humanity and natural factors. The nature of tourist activities continuously changes, which in turn needs improved policies and protection standards. If the varied tourism activities and the diverse consequent impacts are not adequately addressed, they would not be thoroughly considered by international treaties. This could pose unacceptable environmental risks supposedly legally safeguarded by these treaties. Hence, this study elaborated Antarctic tourism impacts in five major groups from a global perspetive. The results are useful for future legislations and implementations.

Keywords: Antarctic tourism; environmental impact; categorization; global perspective

1. INTRODUCTION

The services provided by ecosystems play a vital role for humanity. Ecosystems support the life, supply materials and energy and absorb waste disposals (Daily, 2000) . Thus, maintaining the ecosystems' integrity through appropriate protective means is important. However, ecosystems are complex and these dynamic systems are affected by various temporal and spatial factors (Kariminia, Ahmad, & Hashim, 2012). This makes it difficult to distinguish the levels of stress these systems can endure and the limits of the replacing functions (Gössling, 2002). Meanwhile, the contribution of human activities in the environmental degradation is known. It is believed that these activities reduced the ecosystems' capacity and influenced their services, which in turn, resulted in worrying phenomena such as global warming. Recreation activities, namely tourism undoubtedly contributed to the human-environmental impacts. International treaties e.g. the UN Framework Convention on Climate Change have asked for anticipation, prevention or minimisation of the causes of global climate change and mitigation of its adverse effects (FCCC, 1992). This is more critical on remote destinations such as Antarctica that has fragile and pristine nature. The ways in which these activities impact the environment are the most important step of the related assessment and management processes. The individual influences of tourism activates on local ecosystems have caught research interest, however; due to the interaction relations, environmental degradation aspects should be studied integrally and from a global perspective. Moreover, the role of tourism on environmental stress is not limited to the physical aspect only, but extends to the intangibles (Dyer, Gursoy, Sharma, & Carter, 2007; Nicoletta & Servidio, 2012).

Antarctica is almost thoroughly covered by ice, and the ice sheets supply a large part of earth's fresh water and its thermal mass contributes to the global climate (Kriwoken & Rootes, 2000). Since 1989, almost all sites, including 20 research stations have been visited in the Antarctic Peninsula region (IAATO, 2013). The Antarctic Treaty System consists of international instruments such as 1959 Antarctic Treaty; 1964 Agreed Measures for the Conservation of

Antarctic Fauna and Flora; 1972 Convention on the Conservation of Antarctic Seals, 1980 Convention for the conservation of Antarctic Marine Living Resources; 1991 Protocol on Environmental Protection to the Antarctic Treaty (Madrid Protocol 1991) and other special meetings that made decisions and formed measures and instruments. As the most significant instrument, Madrid Protocol 1991 contained a number of environmental protection instruments. The five annexes, namely Environmental Impact Assessment, Antarctic Fauna and Flora, Waste Disposal Management, Marine Pollution and Area Protection provided a comprehensive measure on the Antarctic region protection (Kriwoken & Rootes, 2000). Nevertheless, an increasing number of academics and parties have raised concern on the Antarctic environment degradations (Bastmeijer & Roura, 2004; Kriwoken & Rootes, 2000).

Historically, the modern era of Antarctic tourism started in the late 1950s. Between 1958 and 1987, the average number of Antarctic tourists did not exceed 1000. However, in a period between 1993 and 1994, this population unprecedentedly exceeded the number of scientists who travelled to this area (Bastmeijer

& Roura, 2004). Despite the limited growth or even gentle decline of visitor pollution in some parts of Antarctic tourism industry, there were close to 35000 visitors in 2012-2013 season; an increase from almost 26000 in 2011-2012 (IAATO, 2013). The increase of passengers on small- and medium-sized vessels and travellers who preferred to experience cruise-only adventure is believed to be the reasons of this jump.

This paper aspires to comprehensively address the environmental impacts of Antarctic tourism on ecosystems from both local and global, and physical and psychological perspectives. In addition, it aims to estimate the scale of these impacts and theoretically assess the role of each for the environmental sustainability. Hence, the Antarctic environmental changes, in both local and global, are firstly reviewed. In the next step, the study concentrated on impacts to which tourism contributes. Both laboratory and field studies were reviewed in these stages. Some parts of the data were presented by previous studies and some parts were reported for other environmental situations or non-systematically explained. The further review, comparison and evaluation shed light on the most critical aspects of the impacts which should be noted in forming the new international treaties.

2. ENVIRONMENTAL CONSEQUENCES OF ANTARCTIC TOURISM

The tourism environmental impacts have occurred locally but added up to a global dimension. From a wide view, these impacts can be categorized into direct and indirect (Gössling, 2002), polluted and non-polluted (Kariminia et al., 2012) and physical and non-physical groups (Nicoletta & Servidio, 2012). Due to the pristine and vulnerable ecosystems in Antarctica, the environmental changes presented by tourism are linked to many factors, namely land use, energy consumption, waste generation, biotic diseases and psychological aspects.

2.1 Discussion

1.1. Changes in land cover and land use

Up to half of the Earth's ice-free surface has been changed by human through land use and land cover alterations (Vitousek, Mooney, Lubchenco, & Melillo, 1997). Land cover change is the physical and biotical alteration at the site while land use change addresses the alterations in the way which humans use the land. Many researchers have reported land alteration as the most important environmental change (Vitousek, 1994; Vitousek et al., 1997) which decisively contributed to increasing the greenhouse gases (GHGs), i.e., CO2 and CH4 (Gössling, 2002). In the tourism industry, infrastructure development is an important part of land alteration. Tourist facilities and infrastructures contributed to a big proportion of impervious surfaces, which in turn, caused more runoff (nutrients, suspended particles, oil and gas) to water bodies (Davies & Cahill, 2000). In the Antarctic continent, the ice-free surface consists only 2% of the total surface (Fig. 1). Furthermore, expeditions were mostly ship-based, and the land adventures generally included short-term ashore visits. In fact, tourism-related constructions in Antarctica are mostly provided for the air base stations and supports. Thus, infrastructure development still remained in a low amount (Lu et al., 2011). Direct local land alterations can be categorized into accommodation establishments (which is limited), land needed for traffic infrastructures and tourist activities and a wider area indirectly affected by the conversions.

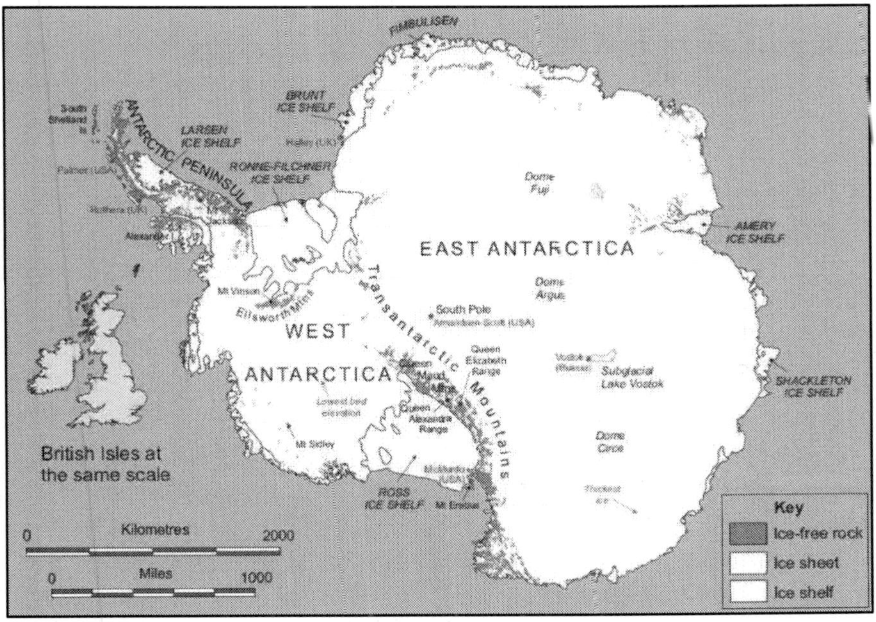

Figure 1. Map of Antarctica (Scentofpine.org)

The accommodation facilities are allocated to the tour operators and the scientific programmes. In terms of traffic infrastructures, airports, marinas, roads and parking contributed to the land conversions. As the only non-governmental permanent tourism air-based facility, the E-base was founded by an NGO in King George Island (Kariminia et al., 2012). It aimed to enhance the public awareness on protecting the Antarctic ecosystem. Furthermore, a Canadian company established a semi-permanent camp at Patriot Hills in 1987, which provided logistic support and organised flights for airborne tourism operations and private expeditions. Although the proportion of covered surface by infrastructures in Antarctica was not considerable, policymakers are concerned with the involved construction and demolition (Liggett, McIntosh, Thompson, Gilbert, & Storey, 2011). In addition, land alterations have indirect impacts such as loss of lands, coastal erosion (Wong, 1998) and sedimentations.

2.2 Energy and material use

A large part of energy use worldwide is generated by fossil fuel consumptions. According to IPCC (2007), the transportation sector significantly relied on fossil resources, particularly petroleum (supplies 95% of the world total energy use by transport). This report argued that in 2004, transportation was responsible for 23% of energy usage-related GHG emissions worldwide. As the most important part of the leisure-related energy use and GHG emissions, transport was responsible for 94% of the total contribution of tourism to global warming (Gössling, 2002).

The Antarctic tourism industry utilised energy for two purposes; transport and destination related. The greater proportion was in transportation as cruise-based travel is the most common expedition. Seven different types of ships are used to ferry tourists: dive boats, expedition ships, icebreakers, motor yachts, Russian ships, sailing vessels and small ships (IAATO, 2013). The fuel consumptions from these vessels result in emissions of GHGs such as carbon dioxide (CO_2), nitrogen oxides (NO_2), sulfur dioxide (SO_2), which in turn contributed to harm biogeochemical cycles (Biesiot & Noorman, 1999) and impact the composition of the atmosphere and biosphere. The most important degradation outcome of these emissions is global warming. In addition to this climate change, the GHG emissions contributed a chain reaction; for instance, rising CO_2 concentrations changed plant physiology (Gössling, 2002). The warmer climate aggregately affected the ecosystems (Sala et al., 2000). Elevated CO_2 and warmer temperatures negatively affected the climate through expansion of higher ecosystems' productivity into the regions Sturm et al., 2001; Tape, Sturm, & Racine, 2006). In fact, terrestrial ecosystems were feedbacks of changing climate as a function of surface energy balance and patterns of sources as well as sinks of atmospheric CO_2. Furthermore, energy use also contributed to changes in other areas of the global environment; for instance, it increased the possibilities for the exchange and dispersion of diseases (Gössling, 2002).

As the most vital resources to humanity, water crisis is expected to exacerbate in the future with regards to climate change and growing human population (Vitousek et al., 1997). As individuals generally used more water while travelling rather than at home (CEED, 1994;

Gössling, 2001), tourism development would result in the increase in overall water use in Antarctica.

Waste production and environmental pollution and disturbance

In general, the role of tourism in environmental pollution and disturbance included travel and destination related aspects. The destination related degradations are categorized into accommodation and activities and involve a wide range of activities, namely construction and maintenance of facilities. The construction and maintenance of facilities generated waste material and energy, which affected the surrounding ecosystem. Although construction in Antarctica is limited, due to the pristine environment, the effect was considerably higher compared to urban areas. The environment pollution could be immediate, such as engine fallout; sewage disposal or gradual such as eutrophication and depleting dissolved oxygen supplies through leaching of nutrients from septic systems into the water body.

Travel

Travel agencies are defined as all the facilities and individuals involved in providing services for tourists. In terms of facilities, large vessels are the highest potential risk as they might have a crash or accident, ground on uncharted rocks, break the ice lands or pollute the water. Operators preferred to use large vessels as small vessel were not economical enough. Liggett et al. (2011) reported twenty-nine accidents and incidents such as damage, aircraft crash; ship grounding and oil spoil recorded in Antarctic area between 1967 and 2003. Surprisingly, almost half of all accidents are accrued during the last 12 years. Although IAATO has provided a swift accessible precautions and assistance, the sinking of MS Explorer in 2007 demonstrated that potential risk of vessels crash still remained (Fig. 2).

Figure 2. Explorer in trouble near South Shetland Islands in 2007 (IAATO, 2011)

In addition, the cruise vessels generally have high potential to pollute the air through engine emissions. It is due to used residual fuels, which have higher contaminants. The annual sulfur emissions by ships can be even higher than that from land (Capaldo, Corbett, Kasibhatla, Fischbeck, & Pandis, 1999). Waste oil is normally generated through leaks from engines, generators and hydraulic systems, and from the fuel filters while conducting maintenance work. Furthermore, toxic chemicals, dry-cleaning wastes, used batteries and paint waste from brush cleaning are possible to occur (Davies & Cahill, 2000). Animals could be exposed to contaminants and discharged wastes. Although the International Convention has prohibited the use and carriage of heavy and intermediate fuel oils for the ships in the Antarctic treaty area for the Prevention of Pollution from Ship since 2009, the concern on the environmental contamination still remains.

The amount of garbage (dry garbage, food waste and disposes) produced by a cruise ship carrying 2700 passengers can exceed a tonne per day (Davies & Cahill, 2000). Illegal dumping of solid waste has been witnessed. For instance, in 1999, Royal Caribbean, the world's second largest cruise line, pled guilty to twenty-one felony counts for dumping oil and chemicals from its cruise ships. Thus, the possibility of both shipboard waste and land generated waste once onshore should be taken into consideration.

Furthermore, airborne travelling could lead to the possibility of air crash, air pollution and wildlife disturbance. After a growth between 1950s and 1970s, the number of tourists frequenting this area via flight seemed to have steadily declined during past few years. However, the potential of crash and degradation related to air based supports exists.

Destination

Tourists inherently tend to visit the most picturesque and wildlife-rich areas with vulnerable ecosystems. According to IAATO (2011), Antarctic tourism currently comprised of eight activities: ship borne expeditions, small boat landing, kayaking, extended walk, station visit, scuba diving, science support and camping (Fig. 3). Site degradation, waste generation and littering, discharging sewage wildlife disturbance and damage to the ice layers could occur with these activities. In addition to the aesthetic issue, littering alters the nutrient compositions and lead to the wildlife disturbance. Sewage contains pathogens, which can contaminate water and affect its quality. The effluent accelerates the growth of algae, which cover the filter-feeding corals and results in hampering their ability to get food. The marine debris, on the other hand, could harm the ecosystems. It aesthetically impacts the coastal areas and ecologically damages the water through gas exchanges between different water surfaces. Moreover, opportunistic organisms could choose debris as its habitat, which could cause changes to the compositions of the ecosystem.

Meanwhile, the new generation of Antarctic tourists who travel individually or in small parties also have a high potential of environmental risk (Fig. 4). The environmental degradation caused by this group would be substantial as they might not have adequate information on the environment they encounter.

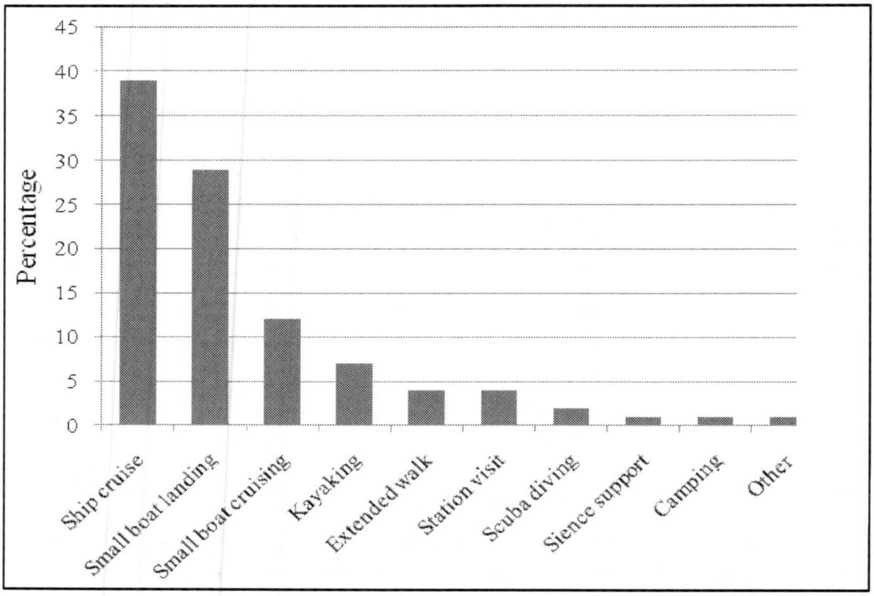

Figure. 3. Distribution of Antarctic tourism activities (IAATO, 2011)

3.5. Fauna and flora exchanges and disease

Enormous exchange of species is a result of human mobility, which leads to homogenization of biota and disruption of natural systems. International commerce, trade in live organisms and transport by tourists are ways to transport species between different environments (Gössling, 2002). The number of new disease-causing organisms discovered in the past three decades has reached thirty (Rodriguez-Garcia, 2001). The exchange and dispersion of diseases through transporting infectious organisms was also a decisive aspect of travel. A growing proportion of travels was nature adventure-based which increased exposure to a greater variety of species and pathogens (Ahlm, Lundberg, Fessé, & Wiström, 1994). Indeed, tourism contributed to the extinction of species both directly and indirectly. The direct ways included voluntary acts such as physical and noise disturbance, littering, and collection of natural objects. However, they can also indirectly disturb the species; for instance, when transporting exotic species, viruses, bacteria, insects or other small organisms. Tourists can transport non-human microbes through their bodies, clothes, animals, goods, food, seeds, etc., which could increase the risk of flora and fauna diseases (Fig. 5).

Figure 4. Individual travellers to Antarctica (IAATO, 2011)

Figure 5 Tourists transport non-native species to the Antarctic environment (IAATO, 2013)

In addition, the facilities related to the tourists' travel and accommodation could lead to the exchange of biota. The accommodation-related facilities, for instance, import plant species alien to the environment. Vessels such as cruise ships are known to transport organism over the long distance of travel to Antarctica. Due to the isolation from other landmasses, Antarctic terrestrial ecosystems currently contain few non-native species. Nevertheless, its indigenous biota is vulnerable to human mediated introductions on non-native species. Hughes, Convey, Maslen, and Smith (2010) reported an incident of the transportation of contaminated soil containing non-native angiosperms, bryophytes, micro-invertebrates, nematodes, fungi, bacteria and seeds through four construction vehicles imported in 2005 by contractors working for the Rothera Research Station (Antarctic peninsula).

There were also rising concerns on the chance of genetic exchange of microbes and evolution of viruses in new environments which in turn would increase the diseases (Goldsmith, 1998). Tourism is believed to contribute to climate change which involve old, new and re-emerging infectious diseases (Kumate, 1997). Climate change was partly caused by the El Niño Southern Oscillation (ENSO) phenomena which led to the reduction in the thickness of the ice layers (IPCC, 2007).

In isolated areas such as Antarctica, tourists exert considerable pressure on animals while closely interacting with them. This reduces the breeding success or threaten them with human pathogens. Antarctic expeditions are usually offered in the austral summer (November to March) as the critical time for the wildlife to breed. For instance, this period is the courting season for penguins, seals being visible on fast ice to establish their breeding territory and penguin chicks start to fledge (IAATO, 2013). The animals would scrounge for the food given by tourists. It could make them accustomed to human food, which would affect their behaviours. Changes in the seal and krill population were one of the results of the marine environment degradation in Antarctic (Wiedenmann, 2010). Penguins show both behavioural and physiological response to visitors which could change their breeding and survival pattern (Bertellotti, D'Amico, & Cejuela, 2013). In addition, the increasing number of tourism communities contributes to changes in the wildlife habitats.

Psychological aspects

The environmental effects of Antarctic tourism are not only limited to the physical aspects. Travelling

to a pristine area also altered the visitors' perception and understanding of the environment as a complex

system of relations between individual, society and nature (Steiner, 1993). This understanding is concerned with environmental knowledge, attitudes and behaviour which can either facilitate or complicate the sustainable processes of the societies (Gössling, 2002). Indeed, there are two paradoxical situations; in one hand, Antarctic tourism promotes visitors' environmental consciousness ecosystem while in the other hand, the proximity to the biota and consumption of the natural resources characterises the personal behaviours of the tourists. Thus, from a psychological perspective, travel to Antarctica alters vironment in two ways:

Environment proximity: Visiting

environment vulnerability. It disrupts their feeling of being responsible to the nature and shrinks it in their eyes. Compared to those who have not this experience, these visitors lack the attachment to nature. The lower responsibility to nature probably increases the consumptions of resources.

Environmental awareness: Exposure to the new representations of the pristine nature fosters the visit

ors' environmental awareness (in addition to the role of education, media or comparison between different environments). For example, visitors perceived wild animals as a representative or symbols of naturalness (O'Rourke, 2000). This alteration in the relations between human and nature increases environmental consciousness, which in turn promotes tendency to travel more.

Fig. 6 depicts the Antarctic ecological tourism impacts in the addressed major categories as well as the defined sub-categories. The severity of the impacts relatively increases from the first to last impact group. Meanwhile, the third group of degradations demonstrates the highest level of direct pollution.

CONCLUSION

This study shed light on the Antarctic tourism related environmental degradations. It aims to elaborate the tourism impacts not only in a local scale, but also from a global perspective. The psychological dimensions are also taken into consideration. The five major fields of these impacts are land alteration, energy and material use, pollution, biotic exchange and change of perception towards the environment.

Land alteration is more allocated to traffic infrastructures, while the accommodation related changes are still limited. The fuel consumption in the transportation results in the GHG emission and contributes to negative impacts on biogeochemical cycles and changes the composition of atmosphere and biosphere. Tourism also increases the usage of fresh water. Environmental pollution is categorized into travel and destination related dimensions. In addition to the wildlife disturbance by through noise pollution, large cruise vessels have run the risk of accident that would pollute the environment. Shipboard and land generated waste contribute to environmental pollution.

Besides the direct disturbances, visitors unintentionally disturb the ecosystem. They transport non- native species through their bodies, clothes and belongings. The cruise shops also carry exotic organisms.

The Antarctic expedition season in summer is a critical time for the wildlife and visitors activities may affect the animals' mating season . Sustainable tourism aims to promote environment-responsible behaviours. Although travelling fosters the knowledge on the environment, it does not necessarily make positive changes in environmental behaviours.

The study demonstrates in detail the susbstantial environmental consequences of tourism in Antarctica. As future study, it is essential to deepen the debate on Antarctic ecosystem protection and address the protection scenarios in the context of global warming.

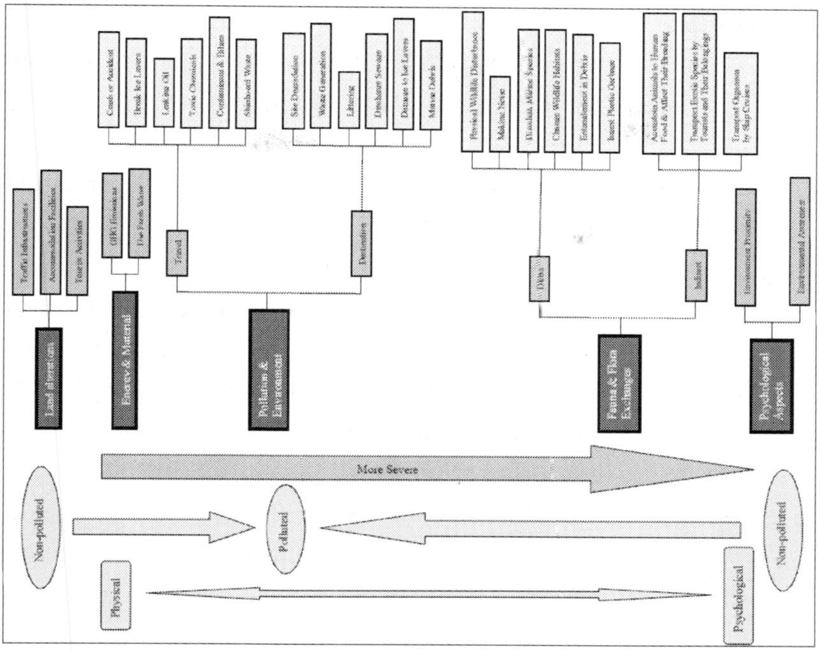

Figure 6. Local and global environmental impacts of Antarctic tourism

ACKNOWLEDGMENT

This study is made possible through research funding by the Ministry of Higher Education Malaysia under the Long Term Research Grant Scheme (Project No: 600-RMI/LRGS 5/3 (4/2011).

REFERENCES

1. Ahlm, C., Lundberg, S., Fessé, K., & Wiström, J. (1994). Health problems and self-medication among Swedish travellers. *Scandinavian journal of infectious diseases,* 26(6), 711-717.
2. Bastmeijer, K., & Roura, R. (2004). Regulating Antarctic tourism and the precautionary principle. *The American Journal of International Law,* 98(4), 763-781.
3. Biesiot, W., & Noorman, K. J. (1999). Energy requirements of household consumption: a case study of The Netherlands. *Ecological economics,* 28(3), 367-383.

4. Capaldo, K., Corbett, J. J., Kasibhatla, P., Fischbeck, P., & Pandis, S. N. (1999). Effects of ship emissions on sulphur cycling and radiative climate forcing over the ocean. *Nature,* 400(6746), 743-746.
5. CEED, U. (1994). A life-cycle analysis of a holiday destination: Seychelles. *British Airways environment report,* 41, 94.
6. Daily, G. C. (2000). Management objectives for the protection of ecosystem services. *Environmental Science & Policy,* 3(6), 333-339.
7. Davies, T., & Cahill, S. (2000). *Environmental implications of the tourism industry*: Resources for the Future.
8. Dyer, P., Gursoy, D., Sharma, B., & Carter, J. (2007). Structural modeling of resident perceptions of tourism and associated development on the Sunshine Coast, Australia. *Tourism Management,* 28(2), 409-422.
9. FCCC, U. (1992). United Nations Framework Convention on Climate Change. *New York, 9.*
10. Goldsmith, M. F. (1998). Health woes grow in shrinking world. *JAMA: The Journal of the American Medical Association,* 279(8), 569-571.
11. Gössling, S. (2001). The consequences of tourism for sustainable water use on a tropical island: Zanzibar, Tanzania. *Journal of Environmental Management,* 61(2), 179-191.
12. Gössling, S. (2002). Global environmental consequences of tourism. *Global environmental change,* 12(4), 283-302.
13. Hughes, K. A., Convey, P., Maslen, N. R., & Smith, R. I. L. (2010). Accidental transfer of non-native soil organisms into Antarctica on construction vehicles. *Biological Invasions,* 12(4), 875-891.
14. IAATO. (2011). Cool Antarctica Retrieved 5 August 2012, from http://www.coolantarctica.com/Antarctica%20fact%20file/science/threats_tourism.htm
15. IAATO. (2013). IAATO Retrieved 14 May 2013, from http://iaato.org/tourism-overview IPCC, A. (2007). Intergovernmental Panel on Climate Change.

16. Kariminia, S., Ahmad, S. S., & Hashim, R. (2012). Assessment of Antarctic Tourism Waste Disposal and Management Strategies towards a Sustainable Ecosystem. *Procedia - Social and Behavioral Sciences,* 68(0), 723-734.

17. Kriwoken, L. K., & Rootes, D. (2000). Tourism on ice: environmental impact assessment of Antarctic tourism. *Impact assessment and project appraisal,* 18(2), 138-150.

18. Kumate, J. (1997). Infectious diseases in the 21st century. *Archives of medical research,* 28(2), 155.

19. Liggett, D., McIntosh, A., Thompson, A., Gilbert, N., & Storey, B. (2011). From frozen continent to tourism hotspot? Five decades of Antarctic tourism development and management, and a glimpse into the future. *Tourism Management,* 32(2), 357-366.

20. Lu, W., Yuan, H., Li, J., Hao, J. J. L., Mi, X., & Ding, Z. (2011). An empirical investigation of construction and demolition waste generation rates in Shenzhen city, South China. *Waste Management,* 31(4), 680-687.

21. Nicoletta, R., & Servidio, R. (2012). Tourists' opinions and their selection of tourism destination images: An affective and motivational evaluation. *Tourism Management Perspectives,* 4(0), 19-27.

22. *Journal of Agricultural and Environmental Ethics,* 13(1-2), 145-165.

23. Rodriguez-Garcia, R. (2001). The health-development link: travel as a public health issue. *Journal of community health,* 26(2), 93-112.

24. Sala, O. E., Chapin, F. S., Armesto, J. J., Berlow, E., Bloomfield, J., Dirzo, R., . . . Kinzig, A. (2000). Global biodiversity scenarios for the year 2100. *Science,* 287(5459), 1770-1774.

25. Scentofpine.org.). Map of Antarctica. Image credit: British Antarctic Survey Retrieved 30 June 2013, from http://scentofpine.org/2008/05/19/faq-antarctica-icemass/

26. Steiner, D. (1993). Human ecology as transdisciplinary science, and science as part of human ecology. *Human Ecology: Fragments of anti-fragmentary views of the world*, 47-76.

27. Sturm, M., Holmgren, J., McFadden, J. P., Liston, G. E., Chapin III, F. S., & Racine, C. H. (2001). Snow-shrub interactions in Arctic tundra: A hypothesis with climatic implications. *Journal of Climate,* 14(3), 336-344.

28. Tape, K., Sturm, M., & Racine, C. (2006). The evidence for shrub expansion in northern Alaska and the Pan Arctic. *Global Change Biology,* 12(4), 686-702.

29. Vitousek, P. M. (1994). Beyond global warming: ecology and global change. *Ecology,* 75(7), 1861-1876.

30. Vitousek, P. M., Mooney, H. A., Lubchenco, J., & Melillo, J. M. (1997). Human domination of Earth's ecosystems. *Science,* 277(5325), 494-499.

31. Wiedenmann, J. (2010). *Implications of climate change for Antarctic krill and their cetacean predators* (Vol. 71).

32. Wong, P. P. (1998). Coastal tourism development in Southeast Asia: relevance and lessons for coastal zone management. *Ocean & Coastal Management,* 38(2), 89-109.

Chapter 11

Environmental Consequences of Antarctic Tourism from a Global Perspective

Shahab Kariminia[a],*Sabarinah Sh. Ahmad[a,*], Rugayah Hashim[b], Zulhabri Ismail[a]

[a] *Faculty of Architecture, Planning and Surveying, Universiti Teknologi MARA, Shah Alam 40450, Malaysia*

[b] *Faculty of Administrative Science and Policy Studies, Universiti Teknologi MARA, Shah Alam 40450,*

ABSTRACT

Antarctica tourism evidently harms its ecosystem in direct and indirect ways. Ecosystems are dynamic and complex systems, which simultaneously depend on various humanity and natural factors. The nature of tourist activities continuously changes, which in turn needs improved policies and protection standards. If the varied tourism activities and the diverse consequent impacts are not adequately addressed, they would not be thoroughly considered by international treaties. This could pose unacceptable environmental risks supposedly legally safeguarded by these treaties. Hence, this study elaborated Antarctic tourism impacts in five major groups from a global perspetive. The results are useful for future legislations and implementations.

Keywords: Antarctic tourism; environmental impact; categorization; global perspective

1. INTRODUCTION

In 2006, the European Council adopted the EU Sustainable Development Strategy. It defines a vision of sustainability in which economic growth, social cohesion and environmental protection are integrated and the needs of the present generation are met without compromising the ability of future generations to meet their own needs (European Council, 2006). European coastal zones can be subjected to intense levels of activities, and many of them face problems of deteriorating natural, socio-economic, and cultural resources. To solve these problems, the European Parlia- ment and the European Council adopted a Recommendation on Integrated Coastal Zone Management (ICZM) in 2002 (CEC, 2002). The European Commission defines ICZM as a dynamic, multi- disciplinary and iterative process designed to promote sustain- able development of coastal zones. Increasing problems in coastal zones and high-ranking political initiatives promoting ICZM have resulted in indicator-based efforts to measure the state of and the progress towards sustainability in coastal zones (Olsen, 2003; Pickaver et al., 2004). Indicators are popular because they provide a simplified view of complex phenomena, quantify information, and make it comparable. Indicators are regarded as important tools in European coastal and maritime policy (Meiner, 2010) and have been used for years to monitor the EU Sustainable Development Strategy.

Given their political usefulness, many coastal indicator sets have been developed on a national (Henocque, 2003; Sarda et al., 2005; Hoffmann, 2009), European (Burbridge, 1997; Van Buuren et al., 2002; Pickaver et al., 2004) and world-wide scale (Ehler, 2003; Olsen, 2003; Belfiore et al., 2006). Many exercises in applying indicator sets (Lescrauwaet et al., 2006; Schernewski et al., 2006; Pickaver, 2009; O'Mahony et al., 2009) and critical evaluations of indicator sets (Breton, 2006; Wallis, 2006; Bell and Morse, 2008) have also taken place. Despite improvements, they revealed several weaknesses, e.g. inadequate recognition and awareness of the indicators' functions, being overly technical and insufficiently oriented towards policy assessment and evaluation and the decision making process (Breton, 2006; Lyytimäki, 2011).

The Guiding Principles for Sustainable Development (CEC, 2005a) mention the coherence between local, regional, national, and global actions, and the review of the EU Sustainable Develop- ment Strategy (CEC, 2005b) points out the importance of the local and regional levels. According to the EU, integrated management of the coastal zone requires strategic, coordinated, and concerted ac- tion at the local and regional levels (CEC, 2002). Thus, coastal municipalities and districts play an important role in sustainable development, and measuring their current state of sustainability and effort is a major task. However, the acceptance of existing in- dicator sets at these administrative levels is very poor. Some rea- sons include complexity, a lack of necessary expertise, data requirements, time costs, results which require interpretation, an uncertain benefit, and a lack of motivation.

Within the project SUSTAIN, a set of indicators has been designed to measure sustainable development in coastal areas on a local and regional level (SUSTAIN partnership, 2012a). The indicator set is linked to a scoring and preference methodology, the DeCyDe tool developed by Isotech Ltd, Cyprus (SUSTAIN partnership, 2012b, Loizidou and Loizides, 2012). The entire methodology can be adjusted to the needs of municipalities and will serve as a decision support and strategic planning tool.

Altogether, we employed nine student groups and one profes- sional expert to apply this indicator set in two Baltic case studies, the German seaside resort Warnemünde and the Lithuanian coastal municipality Neringa. Objectives were to evaluate a) if the indicator set suitably reflects the state of or progress towards sustainability,

b) if it delivers reliable, reproducible results, and c) if it allows for comparisons between different time periods and between different regions. The role of important controlling factors for and the practical relevance of indicator results for planning and manage- ment, as well as future perspectives, will be discussed.

Figure. 1. Location and impressions of the Baltic study sites Warnemünde, Germany and Neringa, Lithuania

2. STUDY SITES AND METHODS

2.2 Study sites Warnemünde and Neringa

The SUSTAIN indicator set was tested in two contrasting coastal study sites in the Baltic Region, the seaside resort Warnemünde in Germany and the coastal municipality of Neringa in Lithuania (Fig. 1).

Warnemünde is part of the city of Rostock, and with 6670 in- habitants (2011) covers an area of 5.9 km^2 and has about 6 km of

coastline. It was founded in the 12th century and remained a small coastal fishery town until the 19th century, when the town was discovered by tourists and seaside holidays at the German Baltic coast became popular. Today, tourism is the major source of in- come, and Warnemünde belongs to the most important of German seaside resorts.

The town provides over 10 000 tourist beds and recorded 313 000 guest arrivals in 2012 and more than 1 000 000 tourist overnight stays (Statistisches Amt Mecklenburg- Vorpommern, 2012). The annual degree of bed capacity uti- lisation is only 27.9%, which reflects the dependency on summer bathing tourism and a relatively short season. A solid pier in War- nemünde protects the entrance of Rostock harbour and causes ongoing accumulation of sand. As a result, the town has a broad sandy beach about 3 km long, and a growing dune belt protects against storm surges. The beach, which has been awarded the Blue Flag, attracts additional visitors from the city of Rostock (204 000 inhabitants in 2011) as well as day visitors from Northern Germany, especially from Berlin. Consequently, the beach is crowded during the summerseason.

Located at the entrance of Rostock harbour and Breitling bay, Warnemünde became an important ship-building location during the 20th century, but the industry has faced a serious decline during the last two decades. After German reunification in 1990 and the resulting political changes in the entire Baltic region, sport- boat and cruise tourism started to grow quickly. In 2012, 181 cruise ships (or 300 000 passengers) visited Warnemünde, making it the most important cruise ship port in Germany. Close to 1 000 sport boats berths are available. Today, fisheries and the small local fish market have only limited economic importance, but are maintained as a cultural heritage and tourist attraction. Parts of the dune belt, the coastal cliffs, and the coastal forests are under nature protection programs.

Neringa municipality is located on the Curonian (Kur_siu˛) Spit ea narrow peninsula, separating the Curonian (Kur_siu˛) Lagoon from the Baltic Sea. It is the longest (about 50 km) municipality of Lithuania at the border to Russia. Neringa was founded in 1961, when the five settlements Nida, Juodkrante, Preila, Pervalka and Alksnyne were joined into one administrative unit. Neringa is part of Kursiu Nerija National Park, a designated HELCOM Baltic Sea Protected Area and a Natura 2000 site. The area is protected as one of the largest and most complex dune habitats in Europe. Moreover, it is an important migratory bird convergence space and known for rare breeding bird species. Forests cover about 83% of total area, but most are protected and used only for recreational purposes (Statistics Lithuania, 2012a). The shoreline between Nida and Juodkrante is relatively stable. Artificial fore-dunes along the Baltic coast protect coastal villages from destructive sand drift.

In 2011, Neringa had 2 570 inhabitants (28.6 inhabitants per km^2) and the population increased by 7.7% during the last decade (Statistics Lithuania, 2012b). Tourism is the major source of income.

In 2012, 69 accommodations hosted 49 456 tourists with 134 786 tourist overnight stays. About 47.4% of the tourists were foreigners. Tourism is concentrated in the summer months, with roughly 72% of overnight stays between June and August (Statistics Lithuania, 2012a). About 12 km out of nearly 50 km of Baltic Sea beaches are used for recreational purposes, have been awarded with the Blue Flag, and possess excellent bathing water quality according the Water Bathing Directive 2006/7/EB. A 53 km bicycle path has been developed, and Nida possesses the only sport boat harbour on the spit.

All land belongs to the state and is only rented to the local population. Agriculture is not allowed in Neringa, and forestry and fisheries account for only about 1% of the total economic turnover. Neringa has a long commercial and cultural tradition in fisheries, but changes during recent years (fish species composition and stock in the Curonian lagoon, fishery restrictions and high real es- tate prices) caused a decline which is considered to be negative for tourism development. Increasing numbers of motorised visitors and infrastructure and urban development coupled with nature protection restrictions have caused ongoing debates in the municipality.

The indicator set

12 partner organisations from across the European Union were involved in SUSTAIN, a 3-year INTERREG IVC programme project partially funded by the European Regional Development Fund. The objective was to create an indicator-based methodology and scoring system which enables local and regional authorities to self- evaluate their sustainability performance for the purpose of improving coastal zone management (SUSTAIN partnership, 2012b). The project followed a bottom-up approach and involved end-users already in the development phase. SUSTAIN provides an indicator set to measure sustainability, with a total of 58 core in- dicators (84 indicators altogether) grouped according to 24 issues, which are then allocated to the four pillars of sustainability: governance, economics, social-wellbeing, and the environment. The first three are represented by five issues, while the last is measured by nine issues. The system is based on indicators that are commonly used and regularly monitored, according to EU legislation. The set of 58 core

indicators should always be applied in study sites, while additional 26 optional indicators allow experts to adapt

the set to local and regional s needs (SUSTAIN partnership, 2012b). The governance issues and indicators are used to measure the consistent management, cohesive policies, guidance, processes, and decisions for good coastal management. Traditionally, indicators to measure governance have proven to be very difficult to define (Bouckaert and Van deWalle, 2003; Ehler, 2003). Therefore,

for this pillar SUSTAIN provides a series of grouped yes e no e don't know questions, with each question regarded as an indicator. Economic performance issues and indicators show whether a strong and sustainable coastal economy is being promoted and supported. Environmental quality performance issues and indicators demonstrate the availability of sustainable environmental practices and theway they are promoted. Social performance issues and indicators measure social unity and resiliance (SUSTAIN partnership, 2012b). Table 1 gives an overview of the core indicators and their allocation to issues and pillars. More detailed descriptions for each indicator and its units are provided in the SUSTAIN partnership (2012a).

After the relevant data is collected and indicator values assigned during the 'indicator application' phase, a moderated stakeholder exercise takes place, which uses matrices to determine the relative importance of the issues and pillars (weighting), which is then combined with the indicator values. Together, both the indicator application and the weighting exercise form the full SUSTAIN methodology, and are included in the DeCyDe tool by Isotech Ltd, Cyprus (Loizidou and Loizides, 2012). We focus on the first part of this methodology, the indicatorapplication.

Indicator application process

The core indicators are mandatory and were used in both study sites, Neringa and Warnemünde. We largely followed the stepwise approach described in SUSTAIN partnership (2012b). First, the relevant data for each core indicator were collected. Second, each indicator was scored using the assessment protocols. The data was then attributed to one of six appropriate classes and converted into class values from 0 to 10 based on predefined ranges. These class values were averaged for each issue and summed to receive a total score for the pillar. If data was imprecise or unavailable, the data was approximated. SUSTAIN provides

EXCEL spread-sheets, which use entered scores to automatically calculate aggregated results for issues and pillars. In a third step, the results would be presented to and discussed with local and regional stakeholders during work- shops. The purpose of this interactive discussion is to evaluate whether the set of indicators both meets local demand and is suf- ficient to provide a realistic picture of the state of sustainability. If not, additional optional indicators can be added to tailor the set to those specific needs. We left this step out of our case study and focused exclusively on scoring core indicators in order to keep the results comparable.

In both study sites, the application exercise was carried out by local postgraduate students (Klaipeda University resp. Rostock University) with varying scientific background. Five groups worked in Neringa in September 2012 (25 students) and four groups in Warnemünde in January 2013 (20 students). Each group had five members: two assessed the environmental indicators, and one person each assessed the social, economic, and governance in- dicators. Each group had several members with good local knowledge. The students received the guidance manuals (SUSTAIN partnership 2012a, 2012b) several days before the application. After an introduction and practical exercises, the groups had about one full day to carry out the indicator application. The exercise was conducted with information publicly available on the Internet complemented by a few telephone interviews with local experts. The students decided by themselves whom to contact and which additional sources to use.

The following day, the groups discussed the scores internally, presented the results to the other students, and provided detailed feedback. The total available time for the application was roughly one working week for one person. The idea was not to apply the most scientifically sound application methodology, but to the test the indicators under the most realistic conditions. The indicators are meant for a self-assessment in municipalities. The educational level and local knowledge of the students, as well as the available time all represent realistic application conditions for typical municipalities. The allocated time was determined from responses from representatives of municipalities and the local tourism sector at a workshop on indicators of sustainability in Warnemünde.

Table 1. The SUSTAIN indicator set: pillars, issues and indicators (modified after SUSTAIN partnership, 2012a).

The SUSTAIN core indicator set: 4 pillars, 22 issues and 58 indicators	
Economics (with 4 issues)	**Indicators** (scoring: 6 classes and 'no data')
1. Economic Opportunity	1. Employment by sector 2. Unemployment rate 3. Gross Value Added
2. Land Use	4. People and assets at risk in in coastal areas
3. Tourism	5. Tourism Intensity 6. Beaches with eco-label awards 7. Recreational boating
4. Transportation	8. Transport usage 9. Volume of port traffic
Environmental Quality (8 issues)	**Indicators** (scoring: 6 classes and 'no data')
1. Air Pollution	1. Air quality
2. Biodiversity & natural resources management	2. Area of land and sea protected by statutory designations 3. Change of condition of coastal and marine habitats and species that have been identified as priorities for conservation
3. Change at the coast	4. Coastal erosion 5. Length of artificially defended coastline
4. Energy & climate change	6. Energy consumption 7. Greenhouse gas emissions 8. Share of renewable energies
5. Land use	9. Area of built-up land
6. Public health and safety	10. Human exposure to harmful noise levels
7. Waste Management	11 Waste production and disposal method 12. Amount of beach litter
8. Water resources and Pollution	13. Quality of bathing water 14. Wastewater treatment 15. Chemical status of coastal waters 16. Ecological status of coastal 17. Hydrocarbon spills
Social well-being (with 5 issues)	**Indicators** (scoring: 6 classes and 'no data')
1. Demography	1. Demographic dependency
2. Equity	2. Poverty
3. Education and training	3. Educational attainment of the population
4. Local and cultural identity	4. Visits to cultural and natural sites 5. Attendance to festivals and public events, organised to strengthen the area's local identity
5 Public Health and Safety	6. Provision of health care services 7. Crime 8. Safety provision
Governance (with 5 issues)	**Indicators** (scoring: 'yes'; 'no'; 'do not know')
1. Policies/strategies for sustainability	1 A sustainable development strategy which includes specific references to the coast and adjacent marine is in place? 2. There is effective political support for the sustainability process? 3. There are integrated, sustainability development plans? 4. The SUSTAIN Issues are covered by relevant policies at the local/regional level? 5. The SUSTAIN Issues are covered by relevant legal instruments at the local/regional level? 6. Guidelines have been produced by national, regional or local governments which advise planning authorities on appropriate sustainable uses of the coastal zone? 7. Strategic Environmental Assessments are used to regularly examine policies, strategies and plans for integration of sustainable activities?
2. Monitoring tools for sustainability	8. Sustainability targets have been set? 9. The sustainability targets are regularly reviewed? 10. There is regular monitoring of the coastal area with respect to the sustainability issues? 11. A report on the State of the Coast has been written with the intention of repeating the exercise every five or ten years? 12. Reviewing and evaluating progress in implementing sustainability criteria is regularly conducted? 13. Assessment of sustainability issues shows a demonstrable trend towards a more sustainable use of coastal and marine resources?
3. Human resources capacity building	14. Local/regional administrations have adequate capacity of staff to deal with sustainability matters? 15. Local/regional administrations have adequate expertise available to deal with sustainability matters? 16. Staff is trained on coastal sustainability matters? 17. All the relevant administrative levels and departments are collectively working on sustainability matters?
4. Implementation of good management practices	18. There is an identifiable point of contact for coastal sustainability matters? 19. Existing instruments are being adapted to deal with sustainability management matters 20. A long-term financial commitment is in place for undertaking initiatives which aim towards sustainability? 21. Integrated programmes on the coast are being carried out that improve the sustainability of the area?
5. Stakeholder involvement/public participation	22. All stakeholders involved in sustainability performance have been identified and are both informed and involved? 23. Partnerships have been established between local authorities and communities for sustainability matters? 24. There is a public participation process involving all necessary stakeholders, including business?

For Warnemünde, a more detailed application also took place. A junior scientist involved in the SUSTAIN project work spent two full working weeks over a period of two months to carry out the application, using Internet, official statistics, literature, and addi- tional phone interviews with local experts.

RESULTS AND DISCUSSION

The indicator application and scoring process

The SUSTAIN indicator set has been selected based on three criteria: relevance to sustainability, data availability, and its readi- ness for field use. The challenges linked with collecting the relevant data for each indicator are indicated in SUSTAIN partnership (2012b) and our experiences confirm several problems, e.g. that the data often is not available from one year, so data from different years has to be used. The consequence is that the indicator appli- cation result does not reflect conditions in municipalities for one reference year, but rather describes the situation during a period of several years. Usually a recent and full data set from only one year was not accessible, and we had to choose a period a few years in the past instead of only one specific year. Therefore, the results are not current. Another problem encountered at both sites was finding data that was specific to the assessed spatial unit. To carry out an indicator application for a traditional and well-defined administrative unit, like a municipality or a district, helps to over- come this problem because the data often is already aggregated with respect to these units. However, in some cases data privacy laws requiring aggregation of data did not allow us to resolve municipal data to a sufficient degree. For Warnemünde, which is a part of the city of Rostock and not a self-governed municipality, suitable aggregated data was often lacking. In our case studies the availability of data was an important problem. Even if data existed, it took effort to find out how and where to access it. The problem of data availability was indicated in other studies as well, e.g. dealing with environmental indicators (Stein et al., 2001), evaluating tourism sustainability (O'Mahony et al., 2009), or discovering in- formation about the local community (Ballinger et al., 2010). One method for to overcome the data availability gap is standard, repeatable, and cost effective information gathering surveys (O'Mahony et al., 2009).

According to SUSTAIN partnership (2012b), 'the approach to score through ranges instead of using precise values, provides the method with flexibility: even data which could not be specifically identified or might be considered imprecise or give just an approximation can be used if identified within a range.' Table 2 shows an example spread-sheet for the issue 'Economic opportu- nity.' In detail, the approach includes several subjective pre- definitions that have significant influence on the results: the defi- nition (boundaries) of the classes, the choice of non-equidistant classes, the definitions of the minimum and maximum of the to- tal range, and the allocation of scores from 0 to 10 to each class. Further, the approach has mathematical weaknesses. If no data is available, the score for an indicator is zero. It is not removed from the calculation but included in the average calculation, reducing the result. Further, indicators that are dependent on each other, like the percentage of employment in primary, secondary and tertiary sectors of the economy (Table 2), are treated as independent in- dicators in the average calculations, causing an overestimate of the indicator 'employment by sector'. Scoring through classes is a simple approach which is easy to understand and allows for the combination of different data (e.g. relative, classified, and numeri- cal data), but includes a problematic loss of information and re- duces the overall quality of the indicator performance. It can hardly be regarded as an advantage in cases where data is uncertain or has to be estimated. Due to these experiences, we thoroughly revised several parts of the scoring spread-sheet.

Reproducibility and reliability of results

Indicator scores are averaged to calculate issue scores, and these are further aggregated into pillar scores. Does aggregation stabilise the results and improve reliability? The average scores for every issue are shown for Warnemünde (Fig. 2) and for Neringa (Fig. 3). For every issue the results between the 4 (5) groups of evaluators differ strongly. The total average over all issues in Warnemünde is five. The averaged minimum scores are two scores lower and the averaged maximum two scores higher than the average. The same is true for Neringa (Fig. 3). The dif- ferences between aggregated results at both issue and pillar levels are very high.

Table 2. Example of the SUSTAIN indicator scoring spread-sheet. Shown are the 3 indicators defining the issue 'Economic Opportunity' as part of the sustainability pillar 'Economic', the scoring units and ranges. Gross Value Added is calculated per sector of economy, explicitly focussing on coastal-dependent activities like fishing, aquaculture, tourism and port activities. The table is modified after SUSTAIN partnership (2012b).

Pillar: Economy								
Issue: Economic Opportunity	Units	Scoring Ranges						
1. Employment by sector	% Primary sector	<10%	10-20%	20-40%	40-60%	60-80%	>80%	No data
		2	4	10	6	4	2	0
	% Secondary sector	<10%	10-20%	20-40%	40-60%	60-80%	>80%	No data
		2	4	10	6	4	2	0
	% Tertiary sector	<10%	10-20%	20-40%	40-60%	60-80%	>80%	No data
		2	4	10	6	4	2	0
2. Unemployment rate	%	>20%	15-20%	10-15%	5-10%	2-5%	<2%	No data
		1	2	4	6	8	10	0
3. Gross Value Added	% of budget for coastal sectors	<5%	5-10%	10-15%	15-20%	20-25%	>25%	No data
		1	2	4	6	8	10	0

What are the consequences of these uncertainties in the indi- cator application methodology? According to the SUSTAIN partnership (2012b), the methodology shall enable municipalities to repeat the exercise at different time intervals in the future, to determine whether or not they are moving towards greater sus- tainability. This repetition is important for a long term assessment (Mc Cool and Stankey, 2004; Breton, 2006; Ballinger et al., 2010). To assess whether repetitions make sense, we compared the present result of the in-depth application in Warnemünde with an appli- cation based on data of the late 1990s. Only a few indicators had different scores for 1990 and today. The systematic changes as re- flected in the aggregated scores are minor when compared to the multiple major methodological uncertainties. First, the SUSTAIN 'scoring through ranges' approach hides small to medium changes, as most data stays in the same class and therefore receives the same score in the present and in the past. For example, the employment in primary, secondary and tertiary sector in a traditional seaside resort like Warnemünde changed only to a very limited degree during the last decade, with the scores remaining in the same classes. It is unlikely that in the future the changes between these three sectors of employment will cause differences in scores, because the classes are relatively broad.

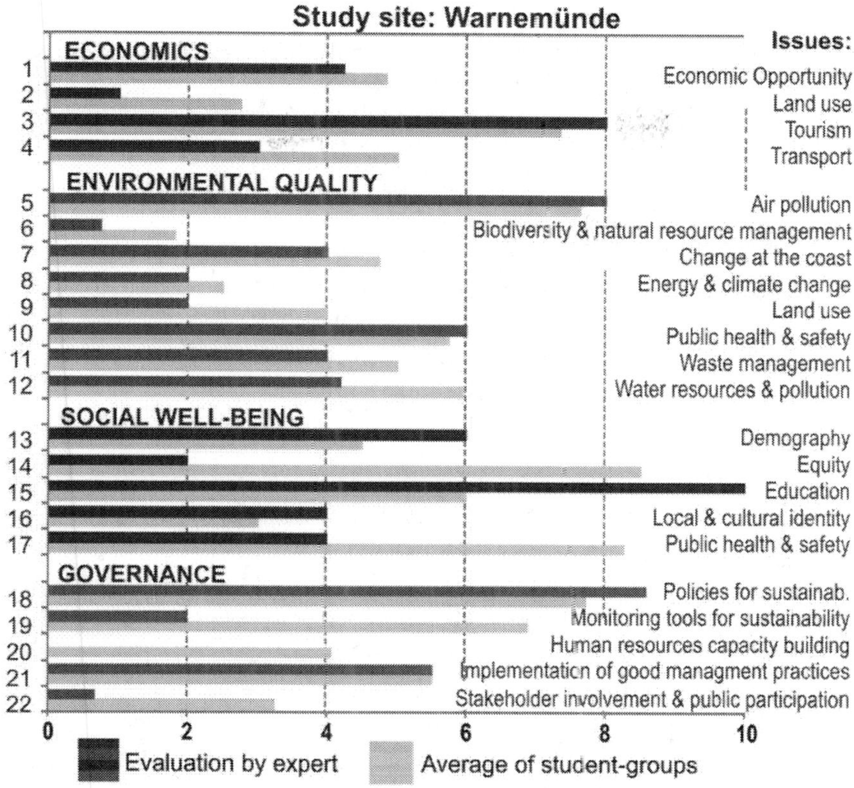

Figure 2. Comparative application of the SUSTAIN core indicator set in the seaside resort Warnemünde. Shown are indicator scores that are aggregated to the level of issues. The issues describe the four pillars of sustainability, namely economics, environmental quality, social well-being and governance. Compared are applications carried out by an expert with average results of 4 student groups

Second, due to data availability, the score always reflects a longer time period rather than a single year, and this reduces dif- ferences between results from two spaces in time. Our experience shows that the indicator set does not allow a reliable comparison of different decades at one study site. Over a long period of several decades, systematic changes might become visible, but only if the quality of data remains stable and the same person carries out the evaluation.

There are several reasons for differences in the results between the groups, including misinterpretations due to insufficient or imprecise indicator descriptions, misunderstandings due to lan- guage barriers

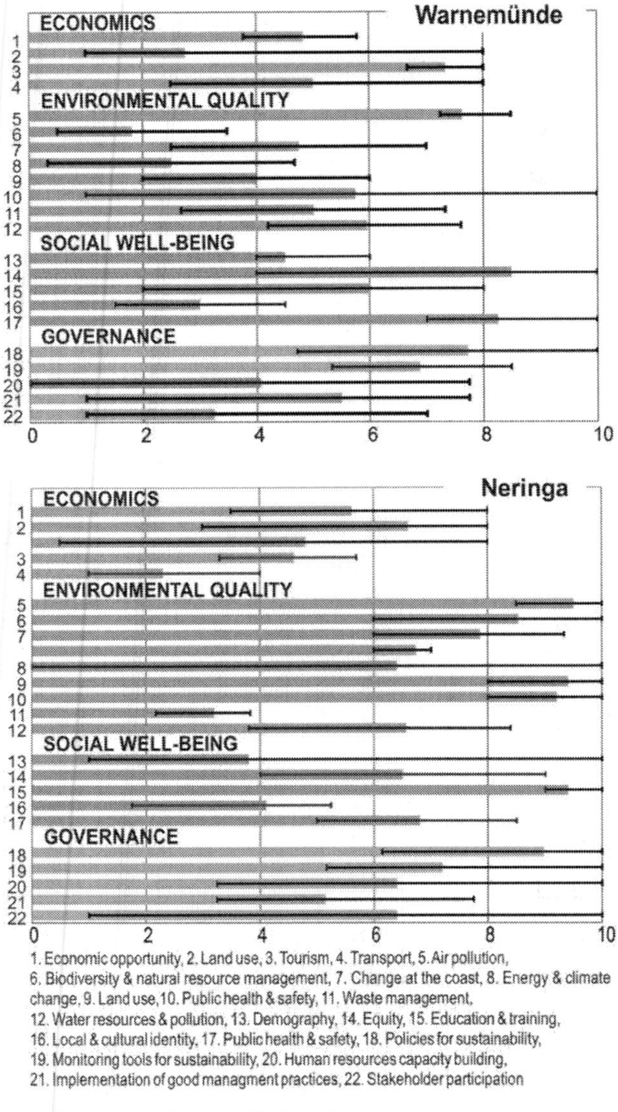

Figure 3. Application of the SUSTAIN core indicator set in the seaside resort Warnemünde, Germany and in the municipality of Neringa, Lithuania. Shown are average, minimum and maximum indicator scores that are aggregated to the level of issues. The issues describe the four pillars of sustainability, namely economics, environmental quality, social well-being and governance. The applications were carried out by 4 (5) student-groups in Germany (Lithuania). During the exercise in Lithuania, the issue 'fisheries and aquaculture' was still part of the pillars economics and environmental quality and were later removed.

(the German and Lithuanian groups used the English indicator description and application manual), and the lack of suitable and sufficiently resolved data combined with the need to estimate certain values. However, subjectivity, perception, and the cultural background of the evaluator also play an important role. This is a known phenomenon even within one country (Ballinger et al., 2010), but become very obvious when groups with very different backgrounds from different countries are involved. Comparative indicator applications between countries involving local evaluators seem hardly reasonable.

The SUSTAIN indicator set has been developed for local municipalities as well as for district and regional authorities, to allow a self-assessment. Local evaluators have the advantage of often bringing good knowledge of the site being assessed and good ac- cess to available local data. However, the lack of experience in in- dicator applications, varying educational levels, and a lack of access to experts all pose problems for the accuracy of the indicator assessment.

Our groups included young scientists with local knowledge who were introduced to the indicators and had the possibility to discuss problems in the group. Discussions within the group very likely stabilised the results and reduced subjectivity. Therefore, it is un- likely that one local person alone and without access to group feedback will produce a more stable, reproducible, and reliable result than our test groups. To what extent does the time available for indicator application affects the quality of the result? Our groups had to base the indi- cator scoring mainly on information and statistics available through the Internet. Official statistics and long-term monitoring programs are considered as the most reliable data sources and therefore very suitable for indicator applications (Hoffmann, 2009). The time available for phone interviews and additional literature search was limited and done only for a few indicators. Compared to this, our in- depth application carried out by a young scientist utilised much more knowledge from local experts, unpublished literature, and planning documents. We did not observe systematic differences in quality between a fast screening, done by a single person within one week (40 h), compared to an in-depth indicator application taking 80 h (two weeks full-time) stretched out over six weeks (Fig. 2). The benefit of investing more time resources into the application process does not improve the result to a degree that it can be regarded as cost-effective. We recommend a restriction of the indicator application to one working week. As mentioned before, the cultural, educational, and disciplinary background of the person carrying out the application has significant impact on the results. On the other hand, local stakeholders with knowledge of the situation and its history as well as suitable contact networks are extremely important (Hoffmann, 2009). However, we also recom- mend the involvement of an experienced external expert in the local application process. A neutral expert can beneficial for a critical view on non-desired development and current local prac- tices (Lyytimäki, 2011).

Spatial, cross-regional comparability

The SUSTAIN indicator sets (core and optional) were developed by a European project involving 12 partners from different Euro- pean countries. The sets should be suitable for all regions in Europe, and a major criterion for the choice of core indicators was their applicability across Europe and the availability of data collected according to various European Directives (SUSTAIN partnership, 2012b). This approach implies that the results allow a comparison of e.g. different municipalities within one country and between different countries.

Does the core indicator set allow cross-regional comparisons between municipalities? To test this, we compared the results from our two study sites (Fig. 3). Both study sites are very different with respect to their local setting, but both are dominated by tourism.

As expected, Neringa receives higher average issue scores for environmental quality. With respect to the issues describing social well-being and economics, both study sites show similar scores. With respect to governance, Neringa has an average of 6.8 compared to 5.5 in Warnemünde. Knowing the situation in both places, the higher value for governance in Neringa and similar scores for economics are surprising. We had expected similar scores for governance and significantly higher scores for economics in Warnemünde. There are several reasons for this surprising result. First, the titles of the issues are sometimes misleading and led to confusion among stakeholders with regard to the indicators describing the issue. The issue 'land use' in the economics pillar, for example, is exclusively described by the indicator 'people and as- sets at risk in coastal areas.' In the environmental quality pillar, the issue 'land use' is described by the indicator 'area of built-up land.' However, both indicators are insufficient to estimate 'land use' as understood in the common sense. The choice of indicators and the terminology caused misconceptions and required revision. Another important reason is the cultural background of the evaluators. The group members seem to intuitively compare the situation in the study site with their experiences from other parts of their country. German students and the expert were more critical of the economic situation in Warnemünde because it belongs to eastern Germany, a region which is known for an under-performing economy. In contrast, the Lithuanian students approved highly of the fast tourism development and improvements in infrastructure in Ner- inga, because this development was perceived as better than the average in Lithuania. Compared to other Lithuanian resorts, like Palanga or Sventoji, Neringa is the most expensive, and attracts Lithuanians with higher income and foreign tourists, especially from Germany. Evaluators might perceive this as an economically sustainable situation and this might influence their evaluation. Possible economic and social risks associated with a short bathing season of only three months and a complete dependency on tourism are perceived as less critical. The cultural and national background seems to play a role for other issues as well, but its effect on the results cannot be quantified based on our data. However, it seems that the indicators and the scoring methodology, which should objectify the evaluation of the state of sustainability, are not able to entirely exclude subjective elements.

Administrative boundaries usually reflect historical or political developments and traditions, which are often similar within one country, for example the criteria how municipalities are defined spatially, but differ between countries. For example, the size and structure of the municipality comprising the study site has imme diate consequences on indicator results. While the urban district Warnemünde is delimited by its administrative boundaries from neighbouring largely rural coastal landscape, these boundaries do not reflect the actual functional relationships along the coast. If the area of Warnemünde included its neighbouring areas, the indicator results would look very different. Largely accidental boundaries have a strong influence on results, which is a problem for inter- regional and international comparisons based on indicators.

Municipalities, districts, and regions show a pattern of hetero- geneous activities and uses rather than a uniform situation. It seems that a heterogeneous study site is more problematic with respect to the application of indicators and the final result will very likely be fuzzier. Therefore, the indicator set should preferably be applied to homogeneous municipalities rather than to larger dis- tricts or regions. Several differences in the issue scores between Neringa and Warnemünde result from different sizes and spatial definitions.

With all these uncertainties, we think that coastal indicators and especially the SUSTAIN core set are not well suited for international comparisons. The strong variability of assessments carried out by different groups for one municipality is present in the end results even for data aggregated to the pillar level (Fig. 4). This high vari- ability would largely conceal differences between different mu- nicipalities, especially on an international level.

Comparisons of municipalities within one country will certainly be more reliable, but it has to take into account that the available data for several indicators (e.g. employment rate) do not differen- tiate on the municipal level but are valid for a region. Municipalities within this region would get the same score for this indicator. Therefore, existing differences between municipalities will not al- ways be sufficiently reflected in the indicator results.

Benefits for practice and local sustainable development

Are the indicators and especially the issues able to reflect the state of sustainable development in municipalities, and does the methodology enable local actors to measure their sustainability effort? The SUSTAIN partnership (2012b) states that 'within coastal zones, there are many hundreds of indicators which purport to give information about sustainability but, in reality, none of them do so e because that is not their purpose e as they are, in general, state- of-the-coast indicators.' The SUSTAIN indicators cover the four pillars of sustainability and are focused on the coast. They can be considered as a step forward, but going through the indicator and issues lists (Table 1) it becomes obvious that most of them have only a weak link to sustainability. However, aggregated to a pillar level they provide insights into the present state of municipalities indicate weaknesses and strengths and, if interpreted correctly, can support decision-making for a more sustainable development.

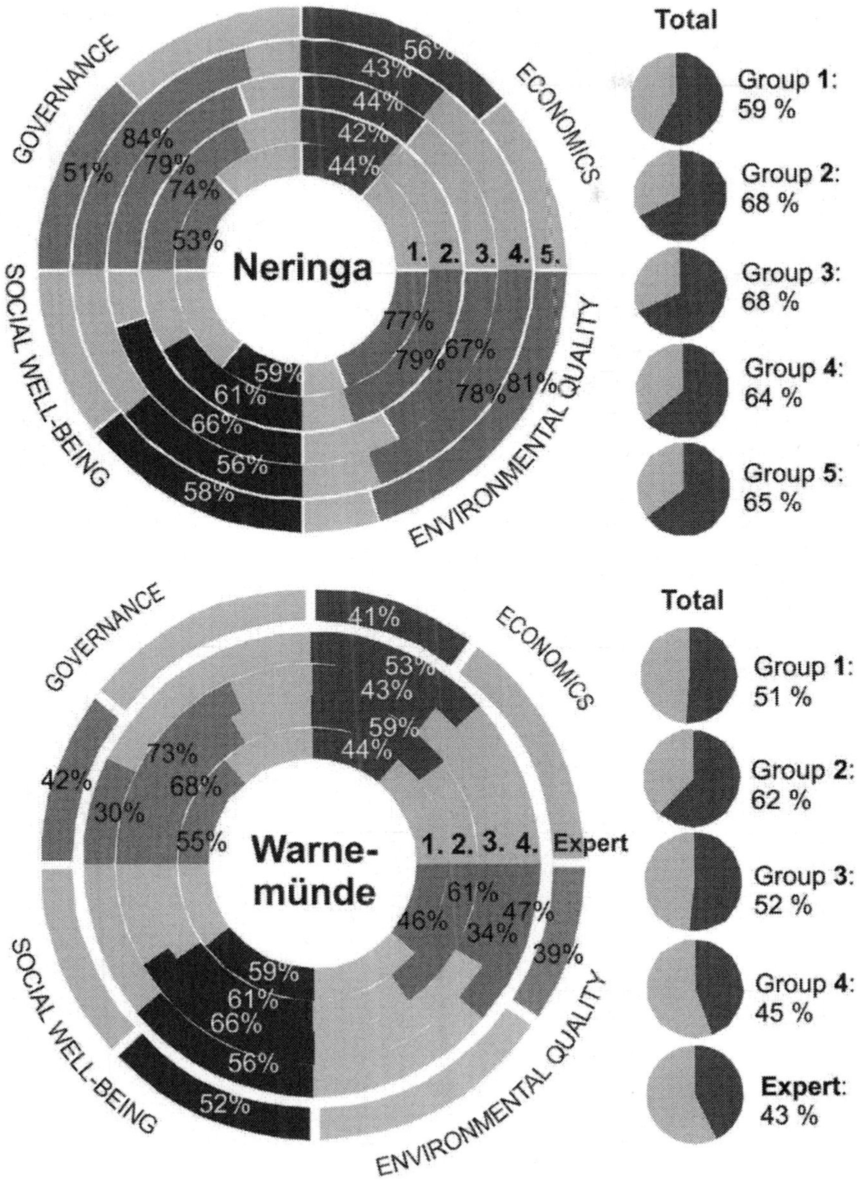

Figure 4. Application of the SUSTAIN core indicator set in the seaside resort Warnemünde, Germany and in the municipality of Neringa, Lithuania. Shown are aggregated results for the 4 pillars (economics, environmental quality, social well-being and governance) as well as total results. The applications were carried out by 4 (5) student-groups in Germany (Lithuania) and one expert.

Indicators and issues only have practical relevance if e.g. mu- nicipalities can make improvements to improve their scores. An improved and more sustainable management has to be reflected in the result, otherwise it is a mere descriptor of the state of the coast indicator. The SUSTAIN optional and core sets include several in- dicators which are beyond local control. Therefore, a revision is necessary to improve their practical relevance.

The aggregated values for pillars and the end results of an application exercise include many uncertainties, and in and of themselves have only very limited practical relevance. The result is less important, than the application process itself. The application process can initiate and guide municipal discussions about sus- tainability. Therefore, the major challenge is the organization, guidance, and maintenance of this process to ensure the participation of relevant decision-makers as well as to involve the public (Mc Cool and Stankey, 2004). Stakeholder engagement and public participation is generally much higher during the early stages of development, particularly during issue identification, yet lacking in long term commitment (Ballinger et al., 2010). Important objectives include raising awareness about what sustainability means and identifying a path towards the creation of a future development vision. The question of how to adapt to climate change challenges is an excellent example of a discussion that could be guided by an indicator application exercise.

The SUSTAIN partnership (2012a) created a core indicator set, which was applied in Warnemünde and Neringa, and additional optional indicators. Optional indicators can be used by municipal- ities if they are relevant and access to the required data is possible. To tailor the indicator set to specific local needs is imperative to ensure a practical value. This approach has to go beyond the SUS- TAIN sets, as municipalities need the freedom to contribute their own, specific additional indicators (Mc Cool and Stankey, 2004). Of course, this approach reduces the regional trans-comparability of the issue and pillar aggregated results even further, and might lead to imbalances in the representation of the four pillars of sustain- ability within one municipality. The wish to compare the status of and progress towards sustainability between regions within one country (Sardã et al., 2005) or even across Europe is a major driver for the development of indicator sets (e.g. Breton, 2006; Lyytimäki, 2011). The indicator set to measure the progress in integrated coastal management (Pickaver et al., 2004), for example, was initiated by EU

Figure 5. Comparison between two indicator systems: the SUSTAIN pillars and issues and the QualityCoast criteria. The colours indicate which issues/criteria could be merged under a joint title. (For interpretation of the references to colour in this figure legend, the reader is referred to the web version of this article.)

DG Environment to get an insight to what extent sustainable management is implemented in different European regions countries and where deficits exist. Comparisons across Europe allow identifying deficits in monitoring and data availability (Breton, 2006). They also include the possibility of learning from other experiences (Moreno-Pires and Fidélis, 2012).

Future perspectives

Indicator sets are hardly used by practitioners on a local level, and there is a risk that this will happen to the SUSTAIN set as well. The practical value of indicator results are hardly visible for municipalities, and using the application process to raise awareness and develop strategies might sound too theoretical for municipal- ities to be willing to allocate time and money. Enabling compari- sons with other municipalities of the region, country, or world, or the idea of a ranking list or a performance map will very likely attract only very few municipalities. This is especially true when the results will be used for promotion and advertisement purposes. Warnemünde, for example, is in keen competition with neigh- bouring seaside resorts, which hampers joint regional advertise- ment programs. It is hard to believe that a publication of indicator results pointing out the strengths and weaknesses of a resort will be welcomed. If indicators are used for internal purposes only, funding will generally be a problem and municipalities will call for external funding schemes (Lyytimäki et al., 2011; Moreno-Pires and Fidélis, 2012).

Our impression is that indicator sets are only attractive and accepted if they ensure an immediate and visible benefit for mu- nicipalities. An existing eco-label, like QualityCoast, could be useful in this respect.

QualityCoast is an international certification programme for sustainable tourism destinations. Since 2007, 125 tourism desti- nations in 23 countries have already been selected for a Qual- ityCoast award. This award includes coastal towns, resorts, and islands (QualityCoast, 2013). The program promises improved awareness of sustainability issues, monitoring strengths and weaknesses, guidance for improvement, transparent information, local publicity, marketing, and promotion (QualityCoast, 2013). QualityCoast offers clear benefits for coastal destinations, and despite focussing on tourism, it uses an indicator system that covers many aspects of sustainability (O'Mahony et al., 2009) and shows many similarities with the SUSTAIN set (Fig. 5). The idea is to technically merge both systems by using the SUSTAIN scoring sheets to increase the motivation to apply the system due to its clear benefits. Municipalities have the short-term benefit that they can directly apply for the QualityCoast label and have the advantage of being able to use the SUSTAIN results to evaluate their state of sustainability and can use it as a policy tool to develop e.g. a sus- tainability strategy.

CONCLUSIONS

The SUSTAIN partnership, (2012a,b) provides sets of core and optional indicators to measure sustainable development in coastal areas on local and regional levels. The indicator set is linked to a scoring and preference methodology and shall serve as a decision support and strategic planning tool. The method- ology is innovative, well documented and, easy to use given the supplementary templates provided for the implementation of the SUSTAIN indicator set. The indicator scores are aggregated into issues and further into the sustainability pillars. Our sys- tematic indicator set applications at two study sites, show that the results have a high variability on all aggregation levels. Reliable spatial comparisons of the state of sustainability in different regions and countries are highly impractical, as are time series applications for single sites. This is a result of short- comings in the availability, quality, and spatial resolution of data, the indicator set and methodology itself, and factors such as human subjectivity. The cultural, national educational and disciplinary background of the evaluator plays an important role as well. For these reasons, the indicator set and the calculation methodology require a thorough revision. However, to our mind, the numerical result of an indicator application is less important than the application process itself. The application process can initiate and guide a discussion about sustainability at the municipal level.

To improve the reproducibility and reliability of indicator re- sults, a homogenous, well defined administrative unit should be selected; municipalities seem most suitable for this. An experi- enced person who is familiar with indicator sets and methodolo- gies should carry out the application in close interaction with local representatives and should serve as a moderator in the discussion workshops. A stepwise process with workshops is necessary to adjust the indicator set to local needs and to discuss the results. 40 work hours (one week) seems sufficient to carry out the basic core- indicator application exercise.

To stimulate the adaptation of this methodology, it must provide clear benefits for municipalities. The initiation of a learning- and awareness-building process and the desire to support strategic planning towards sustainability alone might not be sufficient as motivations.

We recommended the combination of the SUSTAIN methodol- ogy with the QualityCoast labelling system. This could ensure a concrete economic and promotional benefit for municipalities.

ACKNOWLEDGEMENTS

The work has been part-funded by the European Regional Development Fund INTERREG IVC programme project SUSTAIN and by the German Federal Ministry for Education and Research within the project RADOST (01LR0807B).

REFERENCES

1. Ballinger, R., Pickaver, A., Lymbery, G., Ferreria, M., 2010. An evaluation of the implementation of the European ICZM principles. Ocean. Coast Manage 53, 738e749.
2. Belfiore, S., Barbiere, J., Bowen, R., Cicin-Sain, B., Ehler, C., Mageau, C., McDougall, D., Siron, R., 2006. IOC Manuals and Guides No. 46, ICAM Dossier. A Handbook for Measuring the Progress and Outcomes of Integrated Coastal and Ocean Man- agement, vol. 2. UNESCO, Paris.
3. Bell, S., Morse, S., 2008. Sustainability Indicators e Measuring the Immeasurable?, second ed. Earthscan http://www.knovel.com/web/portal/browse/display?_EXT_KNOVEL_DISPLAY_bookid¼2280&VerticalID¼0.
4. Bouckaert, G., Van de Walle, S., 2003. Comparing measures of citizen trust and user satisfaction as indicators of 'good governance': difficulties in linking trust and satisfaction indicators. Int. Rev. Adm. Sci. 69 (3), 329e343.
5. Breton, F., 2006. Report on the use of the ICZM Indicators from the WG-ID. A contribution to the ICZM Evaluation. http://ec.europa.eu/environment/iczm/ pdf/report_wgid.pdf.

6. Burbridge, P., 1997. A generic framework for measuring success in integrated coastal management. Ocean. Coast Manage 37, 175e189.
7. Commission of the Environment European Communities (CEC), 2005b. The 2005 Review of the EU Sustainable Development strategy: Initial Stocktaking and Future Orientations. COM (2005)37 final. Brussels.
8. Commission of the European Communities (CEC), 2002. Recommendation of the European Parliament and of the Council of 30 May 2002 concerning the imple- mentation of Integrated Coastal Zone Management in Europe, 2002/413/EC.
9. Commission of the European Communities (CEC), 2005a. Communication from the Commission to the Council and the European Parliament. Draft Declaration on Guiding Principles for Sustainable Development, COM(2005) 218 final. Brussels.
10. Ehler, C.N., 2003. Indicators to measure governance performance in integrated coastal management. Ocean. Coast Manage 46 (3e4), 335e345.
11. European Council, 2006. Current Sustainable Development Strategy. DOC 10917/06.
12. Henocque, Y., 2003. Development of process indicators for coastal zone manage- ment in France. Ocean. Coast Manage. 46, 363e379.
13. Hoffmann, J., 2009. Indicators for an ICZM. Experience with a problem-oriented approach. J. Coast Conserv. 13, 141e150.
14. Lescrauwaet, A.K., Mees, J., Gilbert, C.R., 2006. State of the Coast of the Southern North Sea: an Indicators-based Approach to Evaluating Sustainable Develop- ment in the Coastal zone of the Southern North sea. In: VLIZ Special Publication, vol. 36. Flanders Marine Institute (VLIZ), Oostende, Belgium.
15. Loizidou, X.I., Loizides, M.I., 2012. DeCyDe: a participatory method for "measuring" sustainability through a friendly, flexible and adjustable [self-assessment?] tool. In: Belpaeme, K., et al. (Eds.), Book of Abstracts. International

Conference Littoral 2012: Coasts of Tomorrow. Kursaal, Oostende, 27e29 November 2012, VLIZ Special Publication, vol. 61, pp. 41e44.

16. Lyytimäki, J., 2011. Indicators and sustainable development strategies: A cross- national overview. In: Lyytimäki, J., Rinne, J., Kautto, P., Assmuth, T. (Eds.), Us- ing indicators to assess sustainable development in the European Union, Finland, Malta and Slovakia. Finnish Environment 4/2011. Finnish Environment Institute, Helsinki.
17. Lyytimäki, J., Kautto, P., Rinne, J., 2011. Barriers and possibilities of indicator use and influence. In: Lyytimäki, J., Rinne, J., Kautto, P., Assmuth, T. (Eds.), Using Indicators to Assess Sustainable Development in the European Union, Finland, Malta and Slovakia. Finnish Environment 4/2011. Finnish Environment Institute, Helsinki.
18. Mc Cool, S.F., Stankey, G.H., 2004. Indicators of sustainability: challenges and oppor- tunities at the interface of science and policy. Environ. Manage. 33 (3), 294e305.
19. Meiner, A., 2010. Integrated maritime policy for the European Union e consoli- dating coastal and marine information to support maritime spatial planning. J. Coast Conserv. 14, 1e11.
20. Moreno-Pires, S., Fidélis, T., 2012. A proposal to explore the role of sustainability indicators in local governance contexts: the case of Palmela, Portugal. Ecol. Indic. 23, 608e615.
21. Olsen, S.B., 2003. Frameworks and indicators for assessing progress in integrated coastal management initiatives. Ocean. Coast Manage. 46, 347e361.
22. O'Mahony, C., Ferreira, M., Fernandez-Palacios, Y., Cummins, V., Haroun, R., 2009. Data availability and accessibility for sustainable tourism: an assessment involving different European coastal tourism destinations. J. Coast Res. 56, 1135e1139.
23. Pickaver, A.H., 2009. Further Testing of the Approved EU Indicator to Measure the Progress in the Implementation of

Integrated Coastal Zone Management in Europe. In: Moksness, E., Dahl, E., Støttrup, J. (Eds.), Integrated Coastal Zone Management. Wiley-Blackwell, Oxford, UK. 10.1002/9781444316285.ch6.

24. Pickaver, A.H., Gilbert, C., Breton, F., 2004. An indicator set to measure the progress in integrated coastal management. Ocean. Coast Manage 47, 449e462.
25. QualityCoast, 2013. http://www.qualitycoast.info.
26. Sardã, R., Avila, C., Mora, J., 2005. A methodological approach to be used in inte- grated coastal zone management processes: the case of the Catalan Coast (Catalonia, Spain). Estuar. Coast Shelf S 62 (3), 427e439.
27. Schernewski, G., Hoffmann, J., Löser, N., Dreisewerd, M., Stavenhagen, P., Grunow, B., 2006. Measuring the Progress and Outcomes of Integrated Coastal and Ocean Management: the German Oder Estuary case study, vol. 21, p. 4. IKZM-Oder Berichte.
28. Statistics Lithuania, 2012a. Lithuania Statistics. http://www.stat.gov.lt/en/ (accessed 07.03.12.).
29. Statistics Lithuania, 2012b. Lithuanian 2011 Population Census in Brief, 48. Statistisches Amt Mecklenburg-Vorpommern, 2012. Tourismus in Mecklenburg-
30. Vorpommern 2012. Statistische Berichte. http://www.statistik-mv.de.
31. Stein, A., Riley, J., Halberg, N., 2001. Issues of scale for environmental indicators. Agr. Ecosyst. Environ. 87 (2), 215e232.
32. SUSTAIN partnership, 2012a. The SUSTAIN Indicator set. A set of easily measurable sustainability indicators. 19 pp. http://www.sustain-eu.net/what_are_we_ doing/sustain_indicator_set.pdf.
33. SUSTAIN partnership, 2012b. Measuring coastal sustainability. A guide for the self-assessment of sustainability using indicators and a means of scoring them. 15 pp. http://www.sustain-

eu.net/what_are_we_doing/measuring_coastal_sustainability.pdf.

34. Van Buuren, J., Smit, T., Poot, G., van Elteren, A., Kamp, O., Künitzer, A., 2002. Testing of Indicators for the Marine and Coastal Environment in Europe, Part 1: Eutrophication and integrated coastal zone management. EEA. Copenhagen, Technical report 84.

35. Wallis, A.M., 2006. Sustainability indicators: is there consensus among stake- holders? Int. J. Environ. Sustain Dev. 5 (3), 287e296

Chapter 12

Fisher and Diver Perceptions of Coral Reef Degradation and Implications for Sustainable Management

Ayana Elizabeth Johnson[a, b, *], **Jeremy B.C. Jackson**[a, c, d]

[a] *Center for Marine Biodiversity and Conservation, Scripps Institution of Oceanography, University of California, San Diego, 9500 Gilman Drive, La Jolla, CA 92083-0202, USA*

[b] *Waitt Institute, 1201 Connecticut Ave. NW, Floor 3, Washington, DC 20036, USA*

[c] *Department of Paleobiology, National Museum of Natural History, PO Box 37012, Washington, DC, USA*

[d] *International Union for Conservation of Nature, 1630 Connecticut Ave. NW, 3rd Floor, Washington, DC, 20009, USA*

ABSTRACT

Fishers and divers are the major resource users of Caribbean coral reefs. On Curaçao and Bonaire, reef condition is good relative to the Caribbean average, but fishes and corals have greatly declined over the last few decades. We interviewed 177 fishers and 211 professional SCUBA divers to assess their views on the extent and causes of degradation. Fishers know fish stocks are severely depleted and declining, whereas divers were aware of declines but had "shifted baselines" and consider the reefs healthy. Fishers and divers differ in perceptions of the causes and appropriate remedies for decline. Fishers generally blame external factors such as changes in climate, currents, or industrial fishing offshore, whereas divers primarily blame overfishing and coastal development. Nevertheless, the great majority of both fishers and divers support more management of both fishing and diving. Thus the social climate is ripe for balanced and strong restrictions on both groups for reef recovery and sustainable use. Exclusion of both fishers and divers from protected areas of significant size around the islands would be a major step forward towards the long-term conservation of reef resources.

Keywords: Artisanal fishing; SCUBA diving; Coral reefs; Reef management; Resource degradation; Shifting baselines

1. INTRODUCTION

Caribbean reefs are severely degraded due to overfishing, coastal development, pollution, and climate change (Jackson, 1997, Hughes et al., 2003, Pandolfi et al., 2005 and Jackson et al., 2014). However, building stakeholder support for managing human impacts is hindered by ignorance of the severity of degradation relative to pristine conditions. This is the problem of "shifting baselines" that plagues understanding of all forms of environmental change, lowering our standards for what healthy ecosystems look like (Pauly, 1995 and Jackson et al., 2011).

Shifting baselines are a problem for industrial fisheries because catch limits are set based upon modern scientific data obtained long after fishing began. Similar problems exist in artisanal fishing communities; more experienced fishers usually consider a greater number of species depleted, note earlier dates for the onset of depletion, and are more likely than less-experienced fishers to notice species declines (Sáenz-Arroyo et al., 2005, Ainsworth et al., 2008, Bunce et al., 2008 and Godoy et al., 2010). Few scientists or recreational divers have ever seen a healthy reef and have little basis for understanding the effects of coastal development and tourism, such as damage to corals by divers (Hawkins et al., 1999 and Barker and Roberts, 2004).

Here we use the Caribbean islands of Curaçao and Bonaire as a case study to explore stakeholders' perceptions of reef fish populations and reef health, and to inform future management. Through extensive interviews of artisanal fishers and professional SCUBA divers, we identified the effects of age, profession, island of residence, and years of experience on (1) expectations for abundance; (2) perceived causes of changes in fish abundance and reef condition; and (3) support for management measures. To our knowledge, this is the first comparative study on the ecological baselines of fishers and divers.

1.1. Local history of fishing and diving and coral reefs

Research was conducted in 2009 and 2010 when Curaçao's resident population was approximately 142,000 versus 13,400 on Bonaire (Central Bureau of Statistics, 2010). Artisanal fishing has occurred on both islands for several hundred years without industrialized fishing or significant exportation. Dive tourism began on both islands in the 1960s. Tourism is the primary economic activity on both islands, especially on Bonaire whose economy is largely dependent on recreational diving (Bouchon et al., 2008). Approximately 44% of Bonaire's tourists buy tags that allow them to dive (over 31,000 tags sold in 2010; Tourism Corporation Bonaire, 2011). Comparable data are unavailable for Curaçao.

The number of fishers on Curaçao has declined from 652 in 1959 (Zaneveld, 1961), to 390 in 1994 (Dilrosun, 2001), to about 200 today (50 full-time and 150 part-time). There are around 30 full-time and 50 part-time fishers on Bonaire; historical numbers are unavailable but anecdotal evidence suggests a several fold decrease. The fishers are largely owner-operators, or working in small groups of one to three crew with a captain. There are around 130 professional SCUBA divers on Curaçao and 120 on Bonaire, spread amongst approximately 40 dive shops.

Average Caribbean coral cover has declined from 35% to 16% since 1970 (Jackson et al., 2014). Coral cover in Curaçao and Bonaire is approximately double this modern Caribbean average but has declined markedly. Fish data are limited but most Caribbean reefs are severely overfished. Wells (1988) reported "overfishing, especially illegal spearfishing," in Curaçao, versus Bonaire where large fish were "far more abundant," and "lacking signs of over-exploitation." More recently, mean fish biomass on the two islands was estimated at 135 g/m^2, with higher values on Bonaire than Curaçao (Sandin et al., 2008; M. Vermeij pers. comm.). Less than 7% of the biomass was apex predators, and no sharks were observed (Sandin et al., 2008). Spearfishing caused sharp declines in parrotfish populations, although Bonaire banned take of parrotfish in 2010 (Steneck et al., 2011). For context, Caribbean fish biomass ranges from as little as 15 g/m^2 in Jamaica to nearly 600 g/m^2 in Cozumel (Newman et al., 2006), with large groupers and snappers virtually absent, and most reefs dominated by smaller species (Pandolfi et al., 2003 and Jackson et al., 2014).

Figure 1. Catch of spearfisher Jose Streder in Curaçao in 1960, (a) a goliath grouper and (b) Nassau and other groupers; and a day's catch in 2009 (c) for one fisher using snorkel (mostly parrotfish) and (d) for two fishers using SCUBA.

Anecdotes from interviewed fishers and divers echo the scientific evidence for sharp declines in fish since the 1980s. In variations on a theme, interviewees express how "the coast is a desert now" compared to what they observed or to stories they have heard. Spear fishers reported having behaved as if "the reef was a supermarket," choosing what they wanted, and catching goliath groupers almost their own size (Fig. 1(a)), or a half dozen Nassau and other groupers (Fig. 1(b)). Large groupers are now rare, spearfishers' catches are dominated by parrotfish (Fig. 1(c)), and two experienced spearfishers using SCUBA were unable to catch a single large grouper or snapper in two dives (Fig. 1(d); A.E.J. pers. obs.). The shift to targeting parrotfish is of particular concern because of their role cleaning the reef of macroalgae that suppress coral recruitment, growth, and survival (Jackson et al., 2014).

Records of hook-and-line catches tell a similar story. In the 1950s, a line fisher in nearshore waters would typically catch 9 to 23 kg of mostly large groupers and snappers (Zaneveld, 1961), but by 1995, 75% of fishers on Curaçao reported catching less than 5 kg per day (unpublished appendix to Debrot and Nagelkerken, 2000). By 2001 the average was down to approximately 4 kg per day (Dilrosun, 2001). There are no time series data of catch landings on either Curaçao or Bonaire.

2. METHODS

In the fall of 2009 on Curaçao and in the spring of 2010 on Bonaire, the first author interviewed 177 full-time and part-time fishers, and 211 professional SCUBA divers, including dive instructors, dive masters, and dive guides. All fishing there can be considered small-scale or artisanal (there is no large-scale industrial fishing or exportation of fish), so here we consider all interviewed fishers as a single group. English–Papiamentu–Dutch translators were used for all fisher interviews, but were unnecessary for divers who all spoke English.

Lack of comprehensive lists of fishers and divers precluded stratified random sampling, so considerable effort was made to include fishers at every port and divers at every shop, to interview fishers and divers of all ages, and to conduct as many interviews as possible. AEJ presented herself as a neutral researcher, unaffiliated with any local group, interested in understanding uses and perceptions of coral reefs and fisheries. At the end of each interview, contact information for additional fishers and divers was requested, in what is termed a snowball sampling strategy (Bernard, 1994).

We acknowledge the possible influence of interviewee wariness of and discomfort about being interviewed. AEJ attempted to combat these effects by (1) letting interviewees set the time and location of interviews to maximize their comfort and convenience, (2) showing photographs of fish (use of images can increase openness and trigger memories; Collier, 1957), (3) using bi-culturally savvy local translators, and (4) conducting interviews in a conversational and casual manner. Interviewees were positively engaged: most gave more detailed responses than requested, none stopped without completing the full interview, many shared fishing and diving stories beyond what was asked, and several later approached AEJ to offer additional information they had forgotten to mention.

Interview questions spanned fishing and diving practices, reef and fisheries status and trends, marine resource management, and demographics. Responses to the questions focused on perceptions of reef resources are presented here (see Appendix A1 for question list). Fishers were questioned about what they catch, and divers were questioned about what they see underwater. Both groups were asked how coral reefs should be managed, including fishing gear restrictions and marine reserves.

We used nine questions about gears and seven questions about reserves to calculate gear scores and reserves scores for each interviewee based on percentage of restrictions they supported. For example, a diver's response that she supported banning use of gill nets would earn one point, whereas a response that she did not would receive zero points, and supporting seven out of nine gear restrictions would earn a score of 77.8. We used responses to these questions about gear and reserves, plus five miscellaneous management questions (i.e., limiting the number of fishers or divers, banning anchoring, banning catch of juveniles, and banning catch of particular species), to calculate an overall conservation score. The highest possible score is 100 (supported all twenty-one restrictions) and the lowest is zero (opposed to all restrictions).

We created a baseline score from responses to questions used to understand interviewees' ecological baselines. For each question, points were based on level of awareness of ecosystem degradation the responses implied. For example, a fisher's response that he catches smaller fish than did previous generations of fishers would earn one point, whereas a response that fish size is the same would earn 0.5, and a response that fish size is bigger would earn zero points. The maximum and minimum possible scores were 7 and zero, with 7 representing an un-shifted baseline. This scoring assumes that fish populations and reef health have been in decline over the last decade, a fact strongly supported by all available evidence.

Fishers and divers were categorized by age: youngest (15–30 years of age), young (31–45), middle (46–60), and old (61–81 years). Altering these age group definitions a few years in either direction did not the change patterns of results. Since there are few professional divers over 60, middle-aged and old divers were combined for analysis. Divers were also categorized by local diving experience (<1<1 year, 1–4 years, 5–10 years, and >10>10 years of diving on that island), and by total diving experience (<10<10 years, 10–19 years, and ≥20≥20 years of global lifetime dive experience). Fishers were categorized by total years of experience (<10<10 years, 10–19 years, 20–29 years, 30–39 years, and >40>40 years).

We compared survey responses between islands and professions, and among age groups and years of experience. For categorical response variables, we used χ^2χ2 goodness-of fit-tests, and where sample sizes were small Fisher's exact tests. For continuous response variables, such as demographic data and best or worst year of fishing, we used one-way analysis of variance (ANOVA).

3. RESULTS

3.1. Interviewee demographics and experience

The 388 interviews involved an estimated 63% of fishers on Curaçao, 65% of fishers on Bonaire, 86% of professional divers on Curaçao, and 83% of professional divers on Bonaire. Interviewed fishers were significantly older than divers (Table 1; statistics for all comparisons are provided in Table 2). Fishers had more experience than divers whether measured in starting age, starting year, or years of experience. Divers on Bonaire had more diving experience than those on Curaçao, both overall and locally. Fishers' families have been fishing for a mean of 3.1 generations, longer than the mean of 1.3 generations for divers. First generation professionals comprised 69% of interviewed divers, compared to only 18% for fishers. Of interviewees with children, only 17% of fishers versus 53% of divers wanted their children to follow in their profession. Just 53% of fishers and 44% of divers considered their profession profitable, and of those who did, about one-third expected that profitability would only last another 10 years or less.

Table 1. Means, with standard errors in parentheses, for demographic information for fishers and divers interviewed on Curaçao and Bonaire.

	Fishers			Divers		
	Curaçao	Bonaire	Overall	Curaçao	Bonaire	Overall
Number of interviewees	126	51	177	112	99	211
Youngest (15–30)	13	13	26	45	32	77
Young (31–45)	34	5	39	52	30	82
Middle (46–60)	58	16	74	13	31	44
Old (61–81)	21	17	38	2	6	8
Mean age	48.6 (1.18)	51.0 (2.71)	49.2 (1.15)	34.3 (0.87)	39.7 (1.23)	36.8 (0.76)
Mean age started fishing or diving	13.7 (0.65)	10.1 (0.92)	12.6 (0.54)	21.9 (0.71)	22.5 (0.92)	22.3 (0.57)
Mean year started fishing or diving	1975 (1.21)	1969 (2.61)	1973 (1.16)	1997 (0.83)	1993 (1.17)	1994 (0.71)
Mean generations of fishers or divers in family	3.0 (0.14)	3.4 (0.17)	3.1 (0.11)	1.3 (0.55)	1.4 (0.53)	1.3 (0.04)
Mean years fishing or diving	34.6 (1.21)	40.8 (2.61)	36.6 (1.16)	12.4 (0.82)	17.0 (1.12)	14.6 (0.70)
Mean years local diving				7.1 (0.09)	10.1 (0.95)	8.5 (0.66)
Mean fishing days per week	3.6 (0.24)	3.8 (0.27)	3.7 (0.17)			
Mean dives per week				8.1 (0.48)	6.2 (0.40)	7.2 (0.33)

Table 2. Statistics for all reported significant differences between groups of interviewees.

Comparison	Statistic
Fishers (compared to divers; Table 1, Fig. 2)	
Older	$F_{1,385} = 84.9$, $p < 0.0001$
Started younger	$F_{1,381} = 148.4$, $p < 0.0001$
Started in an earlier year	$F_{1,382} = 284.3$, $p < 0.0001$
More years of experience	$F_{1,382} = 287.0$, $p < 0.0001$
More generations of their profession in their family	$F_{1,376} = 271.5$, $p < 0.0001$
Less likely to want their children to have their profession	$\chi^2 = 26.9$, $p < 0.0001$
More likely to report a decrease in fish	$\chi^2 = 16.0$, $p < 0.0001$
More likely to perceive fewer fish compared to 10 years ago	$\chi^2 = 15.0$, $p = 0.001$
Younger and less experienced fishers (compared to older and more experienced)	
More likely to report recent years fishing as their best	$F_{3,147} = 11.3$, $F_{3,122} = 9.1$, both $p < 0.0001$
Less likely to report declines in catch	$\chi^2 = 19.8$ and $\chi^2 = 15.49$, both $p < 0.02$
Divers on Bonaire (relative to Curaçao; Table 1, Fig. 2.)	
More years of diving experience	$F_{1,206} = 11.2$, $p = 0.001$
More years of local diving experience	$F_{1,208} = 5.74$, $p = 0.017$
More likely to rate fish populations as abundant or very abundant	$\chi^2 = 17.4$, $p = 0.002$
More likely to have abandoned some dive sites	$\chi^2 = 13.7$, $p < 0.0001$
Divers on Curaçao (but not Bonaire)	
If more local experience, more likely abandoned sites	$\chi^2 = 12.7$, $p = 0.005$
If local experience ≥ 10 years, more likely perceived a decrease in fish	$\chi^2 = 13.9$, $p = 0.031$
Remarks on the absence or decline of large fish more frequently made by	
Divers on Bonaire than Curaçao	$\chi^2 = 14.8$, $p < 0.0001$
Divers with more experience	$\chi^2 = 21.0$, $p < 0.0001$
Older or longer resident divers on Bonaire	$\chi^2 = 10.4$ and $\chi^2 = 14.6$, both $p < 0.02$

More likely to consider the reef unhealthy or very unhealthy	
Older divers	$\chi^2 = 13.0, p = 0.015$
Divers on Bonaire who a longer residents	$\chi^2 = 22.8, p = 0.029$
Divers on Curaçao with more experience	$\chi^2 = 3[illegible].7, p < 0.0001$
Divers over 45 more likely than younger divers to consider reef condition has gotten much worse	$\chi^2 = 17.4, p = 0.008$
More likely to note species absences	
Fishers with more years of fishing experience	$\chi^2 = 11.0, p = 0.012$
Older divers	$\chi^2 = 9.9, p < 0.02$
Divers with more local experience	$\chi^2 = 51.2, p < 0.02$
Divers with more total experience	$\chi^2 = 8.5, p < 0.02$
Causes of degradation (fishers versus divers; Fig. 3)	
Divers more likely to blame fish decline on overfishing, small mesh nets, spearfishing, and coastal development	$\chi^2 = 6.6, 71.5, 25.4, 7.6$, and 57.4 respectively, all $p < 0.04$
Fishers were more likely to blame change in climate, change in currents, SCUBA diving, and the will of God	$\chi^2 = 43.0, 43.0, 7.6$, and 34.0 respectively, all $p < 0.03$
Divers more likely to support gear restrictions, marine reserves, and other assorted conservation measures	all $p < 0.0001$
Fishers more likely to support no diving areas	$p < 0.001$
Divers more likely to support no fishing areas	$p < 0.001$
Baseline scores	
Fisher scores higher than diver scores	$p < 0.0001$
Older divers had lower baseline scores than younger divers	$F_{1,209} = 14.67, p = 0.0003$
Fishers on Bonaire with lower baselines scores were more likely to support gear restrictions and assorted conservation measures	$F_{1,48} = 5.9$ and $F_{1,48} = 7.65$ respectively, both $p < 0.02$
Support for increased management	
More divers than fishers support more fishing management	$\chi^2 = 14.0, p < 0.001$
More divers than fishers support more diving management	$\chi^2 = 6.4, p = 0.04$
More fishers and divers on Curaçao than Bonaire support more fishing management	$\chi^2 = 27.4$, 15.[illegible], both $p < 0.001$
More divers on Curaçao than Bonaire support more diving management	$\chi^2 = 27.7, p < 0.0001$

3.2. Fish catch quantity, abundance, and size

Most fishers (89%) perceived catching fewer fish than previous generations, with no differences in responses between islands, age groups, or years of experience (Fig. 2(a)). Of the few fishers who reported stable catch quantity, two added caveats that they now have a larger boat, go further offshore, and have more advanced gear such as fish finders. Four fishers noted catching less despite better equipment. In contrast, most divers (77%) responded that local fish populations were of average or greater abundance compared to what they would expect a healthy coral reef fish population should look like, with no effect of age group, island or local experience (Fig. 2(d)). Divers on Bonaire were more likely than those on Curaçao to rate their local fish populations as abundant or very abundant (55% versus 33%).

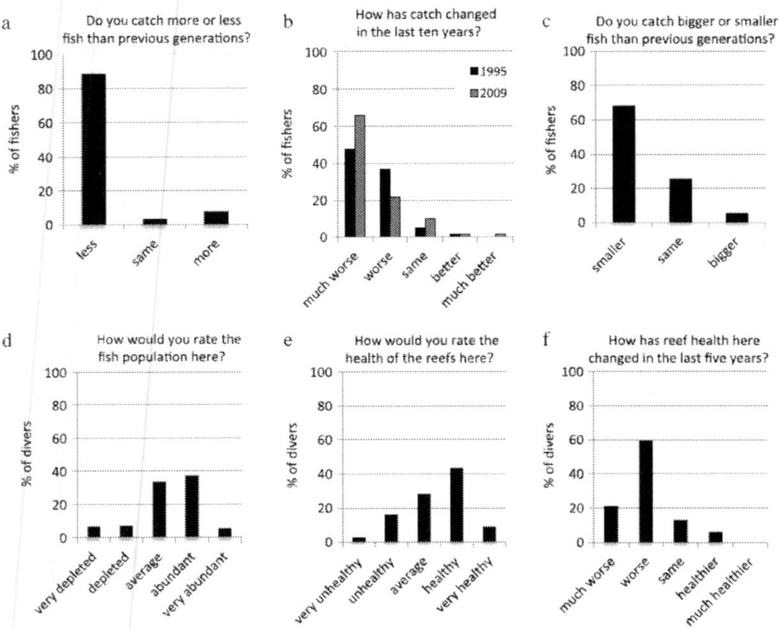

Figure 2. Responses of fishers (a)–(c) and divers (d)–(f) on Curaçao and Bonaire to questions about fish catch and coral reef and fish population health. Question (b) was posed only to fishers with ten or more years of experience, and 1995 data are interpolated from Fig. 2 in Debrot and Nagelkerken (2000). Question (f) was posed only to divers with five or more years of experience.

Interviewees with over five years of experience were asked to compare current fish catch (N=171N=171 fishers) or population size (N=111N=111 divers) to five years prior. More fishers than divers (65% versus 45%) reported a decrease. Younger and less experienced fishers and divers were less likely to report declines. Divers on Curaçao with more than 10 years of local experience (N=26N=26) were more likely to have perceived a decrease than divers with shorter local experience (60% versus 21%). No diver on either island with over 20 years of local experience believed the fish population had increased in the last five years (N=27N=27).

Fishers are somewhat more likely than divers (89% versus 81%) to believe there are fewer fish in the ocean now relative to ten years ago. Neither fisher nor diver responses differed between islands or across age groups or years of experience. When fishers with more than ten years of experience (N=167N=167) were asked how catch had changed in the last ten years, 83% responded that catch was worse or much worse, with no differences in response distributions between islands or age groups (Fig. 2(b)). Compared to fishers interviewed in 1995 (Debrot and Nagelkerken, 2000), a greater portion of our interviewees thought catch had gotten much worse.

The mean year in which fishers reported having the best catches of their lives was 1998, and the mean worst year was 2009. Overall, interviewees most frequently reported their worst year as the year the interview took place. Young and less experienced fishers were more likely to report recent years as their best.

The majority of fishers (69%) reported catching smaller fish than previous generations (Fig. 2(c)), with no effect of island, age group, or experience. Several fishers who reported catching bigger fish than previous generations added the caveat of having better equipment. Five fishers noted catching smaller fish despite better equipment. Comparing his catch with previous generations, a fifteen-year-old fisher remarked, "Now we show fish size horizontally [held hands shoulder-width apart] instead of vertically [held hand at shoulder height off the ground]."

There was no specific question for divers about change in the size of observed fish, yet 47% of divers commented unprompted on the absence or decline of large fish, with more divers on Bonaire than Curaçao remarking on this (58% versus 31%). More experienced divers were more likely to mention a lack of large fish, as were divers on Bonaire who were older or had more local experience. A 44-year-old diver remarked, "I miss the big fish. Some places it's like diving in a saltwater aquarium—all these little fish."

3.3. Reef health, site abandonment, and missing species

Most divers (81%) perceive local reefs to be average or better than average compared to how they think a healthy reef should look, with only 3% of divers considering the reefs very unhealthy (Fig. 2(e)). Older divers were more likely to consider the reefs unhealthy or very unhealthy, as were divers on Bonaire with more local experience, and divers on Curaçao with more overall experience.

Of divers with five or more years of local experience (N=114N=114), 81% consider reef health worse or much worse compared to five years prior (Fig. 2(f)). There were no significant effects of island, local experience or total experience, but on Bonaire, divers over 45 years old were twice as likely as those under 45 to consider reef health much worse (32% versus 16%).

Approximately half of interviewees (48% of fishers and 57% of divers) reported abandoning some fishing or diving locations due to reef or fish population degradation. There were no effects of island, age, or experience on fishers' responses. Some fishers explained they have not abandoned sites because there is nowhere else to go, or their boats or engines are not large enough to go where there are still fish. There were no effects of age or total experience for divers, although (due to storm damage) divers on Bonaire were more likely than those on Curaçao to have abandoned degraded dive sites (70% versus 45%). On Curaçao, likelihood of dive site abandonment increased with local diving experience. Some divers noted their employers require them to continue diving at sites they would otherwise abandon.

An overwhelming 96% of fishers and 94% of divers with over five years of local experience reported that some species they used to catch or see are rare or missing now, with no significant difference between professions. Fishers most frequently mentioned large pelagic fish (i.e., tunas, wahoo, dorado) as greatly reduced. For reef fish, fishers most commonly mentioned reductions in coneys (*Epinephelus fulvus*), groupers (especially Nassau groupers (*Epinephelus striatus*)), and yellowtail snappers (*Ocyurus chrysurus*). Divers most commonly mentioned groupers (again Nassau groupers in particular), green moray eels (*Gymnothorax funebris*), lobster (*Panulirus argus*), and barracuda (*Sphyraena barracuda*).

3.4. Causes of decline and support for management

Fishers and divers differed sharply in perceived causes of fish decline (Fig. 3). Divers were more likely than fishers to blame pollution, overfishing, small mesh nets, spearfishing, and coastal development. Fishers were more likely to blame change in climate, change in currents, SCUBA diving, and the will of God. There were no effects of local or total years of experience. Industrial fishing offshore, sunscreen chemicals scaring away fish and damaging corals, and overfishing bait were the most commonly volunteered additional causes. Compared to Curaçaoan fishers in 1995 (Debrot and Nagelkerken, 2000 and associated unpublished appendix), fishers today appear less concerned with small-mesh nets, spearfishing, and overfishing, and more concerned with the impacts of industrial fishing and SCUBA diving (Fig. 3). Fishers were asked if their fishing reduces the number of fish that will be in the ocean next year or if it damages the reefs; only 11% responded yes to each question. Divers were asked whether diving damages reefs; 94% responded yes.

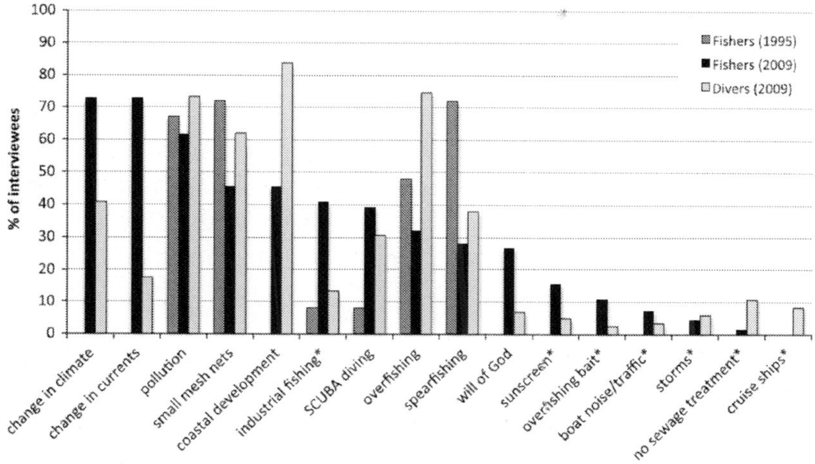

Figure 3. Perceived causes of declines in fish populations according to the present study and data collected in 1995 (Debrot and Nagelkerken, 2000 and the associated unpublished appendix). Asterisks denote factors that interviewees were not questioned about directly, but more than 5% of fishers or divers mentioned as additional contributing factors.

Divers were more likely than fishers to support gear restrictions, marine reserves, and other assorted conservation measures (Table 3). Fisher support of gear restrictions and assorted conservation measures increased with age. Among divers on Bonaire, support for reserves decreased with age. Mean baseline score was higher for fishers than divers; that is, fishers were more likely to report resource degradation. Older divers had lower baseline scores (i.e., less shifted baselines) than younger divers. Fishers on Bonaire with lower baselines scores were more likely to support gear restrictions and assorted conservation measures.

Table 3. Mean scored interviewee responses to questions about reef and fisheries management, where higher numbers represent more conservation-oriented responses, and with standard deviations in parentheses.

	Gear score	Reserve score	Conservation score	Baseline score
Fishers	53.6 (22.0)	36.2 (34.4)	46.5 (18.3)	5.02 (0.1)
Divers	75.0 (22.4)	69.4 (21.6)	72.5 (15.1)	3.22 (0.1)
F-value	89.6	133.4	231.3	162.7

Only 2% of fishers were willing to limit the number of fishers, whereas 34% of divers were willing to limit the number of divers. Sixty-two percent of divers supported closing areas to diving, versus 37% of fishers who supported closing areas to fishing. Unsurprisingly, fishers were more likely than divers to support no diving areas and vice versa for no fishing areas.

Most broadly, the large majority of interviewees (71% of fishers, and 83% of divers) believed there should be more management of fishing, with more fishers and divers on Curaçao holding this viewpoint than their counterparts on Bonaire (79% versus 49%). The majority of interviewees (63% of fishers and 76% of divers) also believed there should be more management of diving, with more divers on Curaçao than Bonaire (91% versus 60%) ascribed to this view, and no significant inter-island difference in fishers' responses.

4. DISCUSSION

4.1. Fisher and diver perceptions are more similar than different

The coral reef ecosystems of Curaçao and Bonaire are in better condition than most places in the Caribbean, but have degraded substantially in recent years due to insufficient management of fishing and coastal development (Jackson et al., 2014). In general, both fishers and divers perceive this decline, although there are some important differences in their perceptions.

Fishers perceive a larger decline in fish populations than do divers. In previous surveys, Curaçaoan fishers unanimously stated that length, weight, and quantity of catches had declined, groupers and snappers were now rarely if ever caught (Schultink and Lindenbergh, 2006), and they had "all but written-off demersal and reef fishing along the southwest coast" because "a total collapse has taken place in the last decade" (Debrot and Nagelkerken, 2000). Remarkably, younger fishers have the same grim perception of fishing conditions as their older peers, and there is no evidence for a shifted baseline in relation to age as documented elsewhere (Sáenz-Arroyo et al., 2005, Ainsworth et al., 2008 and Bunce et al., 2008).

In contrast, divers have a rosier view of the health of fish populations and reefs due to their comparative lack of historical perspective. This shifted baseline is reflected in younger and less experienced divers being less likely to consider the reefs degraded and declining, or to have noticed changes in fish size or species composition. The lack of independence of these variables, and the positive correlation between divers' ages and baseline scores, further buttress this result.

Demographics can explain these differing perceptions of fishers and divers, at least in part. Fishers are generally older, more experienced, and have a long family history of fishing. This has created an oral history, an inter-generational knowledge transfer. The limited timeline of SCUBA diving's existence, and the youth and high turnover of professional divers, likely explain why divers' baselines are shifted. Furthermore, while fishers' incomes are directly related to the kinds, numbers, and sizes of fish they catch, divers' incomes are less immediately coupled to reef health—the number of dive tourists and dive shops on these islands has increased even as the health of the reefs has declined, though we anticipate that trend will not last.

Differences in fisher and diver perceptions of declines in fish populations are accompanied by differences in perceived causes of these trends. Fishers tend to blame large-scale factors, while denying that their fishing has an effect. Divers tend to blame local factors, especially fishing. Thus, it is somewhat surprising that divers are not more pessimistic about the state of reefs and fish populations. Their optimism may reflect a lack of historical ecological knowledge, or a desire to maintain the reputations of these islands as premier dive destinations.

Fishers' responses were consistent between islands, whereas divers' responses differed. Divers on Bonaire were more likely than those on Curaçao to perceive current fish populations as abundant, but also more likely to perceive declines in presence of large fish and reef health. Furthermore, older and more experienced divers on Bonaire were more likely to have perceived these negative trends than newer divers. This makes sense given that, relative to Curaçao, Bonaire has maintained a larger population of large reef fish in shallow water until more recently (Wells, 1988, Sandin et al., 2008 and Steneck et al., 2011). Bonaire also has only 15% of the resident population density per square kilometer as Curaçao, and one quarter the density of tourists (Jackson et al., 2014). Fishing pressure is lower on Bonaire, and there is a stronger conservation ethos. However, there was less support for additional management of fishing and diving on Bonaire, perhaps because considerably more management is already in place there. But taking a step back, the reefs and the perceptions of them are more similar than not on both islands.

4.2. Social underpinnings of the differences in opinions among fishers and divers

Understanding stakeholder perceptions of coral reefs is critical to designing socially and politically palatable management. Differences in perceptions between fishers and divers have policy implications. In an apparent contradiction, fishers are more cognizant than divers of declines but less supportive of management. Beyond the notoriously independent spirit of fishers, this may reflect their belief that they are not causing the problem, which makes restrictions on fishing feel like scapegoating. Fishers have few alternative employment options and strong cultural ties to fishing, and are therefore resistant to having fishing activities constrained, even if they can envision the long-term benefits of such measures. Divers were also reluctant to have their activities constrained, with only one-third of divers supporting limits on the number of divers, despite almost universally acknowledging that diving damages the reefs.

The financial differences between these groups are profound. Concurrently conducted behavioral economics research with the same survey participants found fishers to be less financially patient than divers because they cannot afford to wait for the long-term benefits of sustainable management and marine protected areas (Johnson and Saunders, 2014). Thus it is critical to carefully manage the transition from existing fishery management regimes to more sustainable ones to ease the burden on fishers.

Racial differences are also relevant. Divers are overwhelmingly white and foreign, while fishers are overwhelmingly black and Antillean, and the sting of colonialism is often barely below the surface. Thus, a balanced, nuanced management approach is needed. While tourism is the main economic driver, fishing is important socioculturally and to a different demographic. Age differences are also important. Older fishers are more supportive of management than younger fishers, presumably because they have witnessed the decline of their fishery first hand. In contrast, younger divers were more supportive of reserves than were older divers, implying a potential tipping point in the professional consensus of that sector as older divers retire.

4.3. Implications for management

Despite all these differences between fishers and divers, both groups largely agree on two critically important matters: health of the reefs and fisheries is severely and increasingly threatened, and more and better management is needed to turn this situation around. Moreover, only around half of interviewees even consider their profession profitable, thus, it is unsurprising that the number of fishers on Curaçao and Bonaire has rapidly declined, and they do not want their children to be fishers. This gives the governments a mandate to more strongly manage ocean resources because both groups agree there is big problem, and their livelihoods depend on it being solved. The question is which management interventions can ensure fishers can continue to fish while dive tourism can also thrive.

In seeking solutions, we must note that both sectors cause significant damage to the environment. Fishers tend to get picked on because their impacts are so immediate and obvious, but impacts of tourism can be just as bad (Jackson et al., 2014). The damage SCUBA divers cause by kicking and touching reefs is as much of a concern as fishers depleting key fish species, especially given the explosion in the numbers of divers (Hawkins et al., 1999 and Barker and Roberts, 2004). For example, an interviewed dive instructor described how conflicted he is about pointing out seahorses to tourists, because while that leads to bigger tips it also creates a halo of damage as tourists all try to get close and take photos. Research strongly supports his observations (Uyarra and Côté, 2007). This sort of diver damage is a global problem that has spawned the Greenfins movement (http://www.greenfins.net/) to train professional divers to minimize impacts of their clients. However, success of this and other voluntary diver education programs has been modest, and restrictions on divers need to be formally incorporated into reef management. Thus, both sectors need to be regulated, and we must recognize that there is a carrying capacity for both.

4.4. Recommendations for management

Bonaire currently has a few small marine reserves, including two closed to diving, and a suite of fishing gear restrictions. Curaçao also has gear restrictions but has not established any reserves closed to fishing or diving. Despite the dedicated efforts of several well-informed and devoted civil servants, current management is inadequate to stem declines in reef health around both islands.

Several policy actions are essential to providing the foundation for long-term sustainable use of coral reefs around Curaçao and Bonaire. First, establish large no-take marine reserves that are closed to both fishing and diving. These reserves should be established at the most ecologically valuable sites, not places no one is keen on using anyway. Marine reserves can help combat the shifting baselines problem by restoring areas and opening a window into the past (Bohnsack, 2003 and Knowlton and Jackson, 2008). More broadly, ocean zoning based upon intensive community consultation can provide stakeholders with priority areas for their use, thereby minimizing conflicts and creating conservation zones (Johnson, 2014).

Second, limit the number of fishers and divers. The large proportion of elderly fishers, and overall decline in number of fishers, provide resource managers an opportunity to limit their numbers going forward, potentially with low social resistance since few young people are pursuing the profession. Following the law of supply and demand, limiting the number of divers could create a higher-value boutique tourism market.

Third, educate stakeholders about historical ecosystem baselines to ensure long-term support for strong management measures. Divers could benefit from interactions with elderly fishers who have important historical knowledge (Johannes et al., 2000). Fishers could benefit from learning more about the ecological impacts of fishing. Tourists could benefit from more deeply understanding how severely they damage reefs. And, scientists could benefit from learning how to ensure their coral reef research produces policy-relevant results. Although there is an increasing trend toward researchers including more stakeholder data, it is not enough, and many scientists neglect to make policy recommendations based on the results of their research (Johnson et al., 2013).

Fourth, ease the transition to more sustainable use of coral reef resources. Buy-outs of both fishers and divers could hasten the process. It may be appropriate to establish buyback programs for types of fishing gear or dive shops that are phased out. Heightened recreational dive fees could be used to offset short-term losses and strengthen protections. Many of the would-be next generation of fishers and divers could also be trained to help monitor and enforce protected areas.

Taking these actions in concert would increase expectations of ecosystem abundance and productivity, facilitate development of appropriately ambitious ecosystem recovery goals, and reduce the strain of the transition to sustainable management. The economic and cultural futures of fishing and diving depend on such measures, and strong, science-based management.

ACKNOWLEDGMENTS

We thank the fishers and divers who generously shared their time and knowledge, and wonderful translators K. Brown, M. Brown, J. Vollenbregt, D. Abraham, C. Dovale, H. Liber, and M. Chirino. F Dilrosun, R. de León, E. Beukenboom provided generous logistical support and historical perspective. S. Sandin, T. Groves, L. Levin, and M. Lauer provided valuable comments. A.E.J. received funding from NSF Graduate Research, NSF IGERT Grant No. 0333444, Switzer Foundation, and AAUW fellowships, and Scripps Institution of Oceanography.

REFERENCES

1. Ainsworth, C.H., Pitcher, T.J., Rotinsulu, C., 2008. Evidence of fishery depletions and shifting cognitive baselines in Eastern Indonesia. Biol. Cons. 141, 848–859.
2. Barker, N.H.L., Roberts, C.M., 2004. Scuba diver behaviour and the management of diving impacts on coral reefs. Biol. Cons. 120, 481–489.
3. Bernard, H.R., 1994. Research Methods in Anthropology: Qualitative and Quantitative Approaches. Sage Publications, Thousand Oaks, CA, p. 585. Bohnsack, J.A., 2003. Shifting baselines, marine reserves, and Leopold's biotic ethic. Gulf Caribbean Res. 14, 1–7.

4. Bouchon, C.P., et al., 2008. Status of coral reefs of the lesser antilles: the French West Indies, the Netherlands Antilles, Anguilla, Antigua, Grenada, Trinidad and Tobago. In: Wilkinson, C. (Ed.), Status of Coral Reefs of the World: 2008. Global Coral Reef Monitoring Network and Reef and Rainforest Research Center, Townsville, Australia.
5. Bunce, M., Rodwell, L.D., Gibb, R., Mee, L., 2008. Shifting baselines in fishers' perceptions of island reef fishery degradation. Ocean Coast. Manag. 51, 285–302.
6. Central Bureau of Statistics,, 2010. Statistical Yearbook of the Netherlands Antilles. Netherlands Antilles. Willemstad, Curaçao. Collier, J., 1957. Photography in anthropology: A report on two experiments. Am. Anthropol. 59, 843–895.
7. Debrot, A.O., Nagelkerken, I., 2000. User perceptions on coastal resource state and management options in Curacao. Rev. Biol. Trop. 48, 95–106. Dilrosun, F., 2001. Progress Report Curaçao Fishery Monitoring Program. Dienst Landbouw, Veeteelt & Visserij, Willemstad, p. 13.
8. Godoy, N., Gelcich, S., Vasquez, J.A., Castilla, J.C., 2010. Spearfishing to depletion: evidence from temperate reef fishes in Chile. Ecol. Appl. 20, 1504–1511.
9. Hawkins, J.P., Roberts, C.M., Van't Hof, T., De Meyer, K., Tratalos, J., Aldam, C., 1999. Effects of recreational scuba diving on Caribbean coral and fish communities. Conserv. Biol. 13, 888–897.
10. Hughes, T.P., et al., 2003. Climate change, human impacts, and the resilience of coral reefs. Science 301, 929–933.
11. Jackson, J.B.C., 1997. Reefs since Columbus. Coral Reefs 16, supplement: S23–S32.
12. Jackson, J.B.C., Donovan, M.K., Cramer, K.L., Lam, V.V. (Eds.), 2014. Status and trends of Caribbean coral reefs: 1970–2012. Global Coral Reef Monitoring Network, IUCN, Gland, Switzerland.
13. Jackson, J.B.C., E., Sala, K., Alexander (Eds.), 2011. Shifting Baselines: The Past and the Future of Ocean Fisheries. Island Press, USA.
14. Johannes, R.E., Freeman, M.M.R., Hamilton, R.J., 2000. Ignore fishers' knowledge and miss the boat. Fish Fish. 1, 257–271.

15. Johnson, A.E., 2014. Small Caribbean island shows bold ocean leadership: Barbuda overhauls reef and fisheries management for sustainability. National Geographic Voices, Washington, DC, USA. Available from: http://voices.nationalgeographic.com/2014/08/13/barbuda-overhauls-ocean-management- for-sustainability/ (accessed April 2015).
16. Johnson, A.E., Cinner, J., Hardt, M., Jacquet, J., McClanahan, T., Sanchirico, J., 2013. Trends current understanding and future research priorities for artisanal coral reef fisheries research. Fish Fish. 14 (3), 281–292.
17. Johnson, A.E., Saunders, D.K., 2014. Time preferences and the management of coral reef fisheries. Ecol. Econ. 100, 130–139. Knowlton, N., Jackson, J.B.C., 2008. Shifting baselines, local impacts, and global change on coral reefs. PLos Biol. 6, 215–220.
18. Newman, M., Paredes, G., Sala, E., Jackson, J., 2006. Structure of Caribbean coral reef communities across a large gradient of fish biomass. Ecol. Lett. 9, 1216–1227.
19. Pandolfi, J.M., et al., 2003. Global trajectories of the long-term decline of coral reef ecosystems. Science 301, 955–958.
20. Pandolfi, J.M., et al., 2005. Are U.S. coral reefs on the slippery slope to slime? Science 307, 1725–1726. Pauly, D., 1995. Anecdotes and the shifting baseline syndrome of fisheries. Trends Ecol. Evol. 10, 430.
21. Sáenz-Arroyo, A., Roberts, C.M., Torre, J., Cariño-Olvera, M., Enríquez-Andrade, R.R., 2005. Rapidly shifting environmental baselines among fishers of the Gulf of California. Proc. R. Soc. B 272, 1957–1962.
22. Sandin, S.T.A., Sampayo, E.M., Vermeij, M.J.A., 2008. Coral reef fish and benthic community structure of Bonaire and Curacao, Netherlands Antilles. Caribbean J. Sci. 44, 137–144.
23. Schultink, R., Lindenbergh, S., 2006. Intensiviteit van de visserij op Curaçao.
24. Steneck, R.S., Arnold, S., DeBey, H., 2011. Status and trends of Bonaire's coral reefs: Cause for grave concerns. Tourism Corporation Bonaire,, 2011. Bonaire Tourism: Annual Statistics Report 2010.

25. Uyarra, M.C., Côté, I.M., 2007. The quest for cryptic creatures: Impacts of species-focused recreational diving on corals. Biol. Cons. 136, 77–84.

26. Wells, S.M. (Ed.), 1988. Coral Reefs of the World. Volume 1: Atlantic and Eastern Pacific. UNEP/IUCN, IUCN, Gland, Switzerland and Cambridge, U.K./UNEP, Nairobi, Kenya.

27. Zaneveld, J.S., 1961. The fishery resources and the fishery industries of the Netherlands Antilles. Proc. Gulf Caribbean Fish. Inst. 14, 137–171.

Index